United States
Department of
Agriculture

Forest Service

Agriculture
Handbook
No. 680

June 2012

Forest Nursery Pests

Michelle M. Cram
Michelle S. Frank
Katy M. Mallams
Technical Coordinators

This is a revision of *Forest Nursery Pests*, Agriculture Handbook No. 680, that was issued in December 1989.

Acknowledgments

This edition of *Forest Nursery Pests*, Agriculture Handbook No. 680, was made possible by the work of many people from around the country. Contributing authors include U.S. Department of Agriculture, Forest Service and Agricultural Research Service entomologists and pathologists, university professors and researchers, State extension specialists, consultants, and plant pathologists and horticulturists in private industry. We thank all of these people for volunteering their time and expertise. We also thank the many individuals and agencies that allowed the use of their photographs.

Sonja Beavers of the Office of Communication in the Forest Service Washington Office shepherded the manuscript through the publication and printing process. Diane Haase, Forest Service Western Nursery Specialist, arranged for publishing on the Reforestation, Nurseries, and Genetics Resources Web site (http://www.rngr.net).

We thank the regional directors of the Forest Service Forest Health Protection for supporting this project and Doug Daoust, Director of Forest Health Protection in the Pacific Northwest Region, in particular, for making the commitment to fund publication and printing.

We also thank the authors of the previous forest nursery pest and disease handbooks. Without their work, this publication would not have been possible. To them, and to all those who participated in the production of this handbook, thank you.

Michelle M. Cram
Michelle S. Frank
Katy M. Mallams

Contents

Diagnosis of Pest Problems
Alex C. Mangini 1

Integrated Nursery Pest Management
R. Kasten Dumroese 5

Evaluating Damage to Nursery Crops
Katy M. Mallams and Tom Starkey 13

Soil-Pest Relationships
Jerry E. Weiland 16

Mycorrhizae in Forest Tree Nurseries
Michelle M. Cram and R. Kasten Dumroese 20

Pesticide Regulations
John W. Taylor, Jr. 24

Conifer Diseases

1. **Brown Spot Needle Blight**
   Scott A. Enebak and Tom Starkey 28

2. **Cylindrocarpon Root Disease**
   Robert L. James 31

3. **Diplodia Shoot Blight, Canker, and Collar Rot**
   Glen R. Stanosz 33

4. **Dothistroma Needle Blight**
   Glen R. Stanosz 36

5. **Eastern and Western Gall Rusts**
   Michael E. Ostry and Jennifer Juzwik 39

6. **Fusiform Rust**
   Scott A. Enebak and Tom Starkey 41

7. **Larch Needle Cast**
   Katy M. Mallams 44

8. **Needle Cast Diseases of Pines**
   Katy M. Mallams 47

9. **Passalora Blight**
   Charles S. Hodges and Michelle M. Cram 50

10. **Pestalotiopsis Foliage Blight**
    Scott A. Enebak 52

11. **Phoma Blight**
    Robert L. James 54

Contents

12. **Phomopsis Blight**
    Edward L. Barnard 56

13. **Phomopsis Canker**
    Katy M. Mallams 58

14. **Pitch Canker of Pines**
    Scott A. Enebak, Tom Starkey, and Tom Gordon 60

15. **Rhizoctonia Blight of Southern Pines**
    Tom Starkey and Scott A. Enebak 63

16. **Scleroderris Canker**
    Michael E. Ostry and Jennifer Juzwik 66

17. **Sirococcus Shoot Blight**
    Glen R. Stanosz 68

18. **Snow Molds of Conifers**
    Jill D. Pokorny 71

19. **White Pine Blister Rust**
    Lee E. Riley and Judith F. Danielson 74

Conifer Insects

20. **Cranberry Girdler**
    Art Antonelli 80

21. **Pine Tip Moths**
    John T. Nowak 82

22. **Sawflies**
    John T. Nowak 84

Hardwood Diseases

23. **Anthracnose**
    Jill D. Pokorny 88

24. **Leaf Spots and Blights**
    Michelle M. Cram and Will R. Littke 91

25. **Marssonina Blight**
    Michael E. Ostry and Jennifer Juzwik 94

26. **Poplar Cankers**
    Michael E. Ostry and Jennifer Juzwik 96

27. **Poplar Leaf Rusts**
    Michael E. Ostry and Jennifer Juzwik 98

28. **Powdery Mildew**
    Michelle M. Cram and Glen R. Stanosz 100

Hardwood Insects

29. Cottonwood Borers
Forrest L. Oliveria and James D. Solomon 104

30. Cottonwood Leaf Beetle
Forrest L. Oliveria and James D. Solomon 107

Conifer and Hardwood Diseases

31. Charcoal and Black Root Rots
Edward L. Barnard 110

32. Cylindrocladium Diseases
Edward L. Barnard and Jennifer Juzwik 112

33. Damping-Off
Robert L. James 115

34. Fusarium Root and Stem Diseases
Robert L. James 117

35. Gray Mold
Diane L. Haase and Michael Taylor 121

36. Nematodes
Stephen W. Fraedrich and Michelle M. Cram 123

37. Phytophthora Root Rot
Michelle M. Cram and Everett M. Hansen 126

38. Pythium Root Rot
Jerry E. Weiland 129

39. Seed Fungi
Stephen W. Fraedrich and Michelle M. Cram 132

40. Sudden Oak Death
Gary Chastagner, Marianne Elliott, and Kathleen McKeever 135

41. Yellows or Chlorosis
Tom Starkey 138

Conifer and Hardwood Insects

42. Aphids
Art Antonelli 142

43. Cutworms
Michael E. Ostry and Jennifer Juzwik 144

44. Field and Short-Tailed Crickets
Coleman Doggett 146

Contents

45. **Fungus Gnats**
    Art Antonelli 148

46. **Lesser Cornstalk Borer**
    Wayne N. Dixon and Albert E. Mayfield, III 150

47. **Mole Crickets**
    Coleman Doggett 153

48. **Plant Bugs**
    David B. South 155

49. **Seed and Cone Insects**
    Alex C. Mangini 158

50. **Mites**
    Alex C. Mangini 163

51. **Weevils**
    Art Antonelli 167

52. **White Grubs**
    Albert E. Mayfield, III 170

Miscellaneous Pest Problems

53. **Animal Damage**
    David B. South 174

54. **Environmental and Mechanical Damage**
    Michelle M. Cram 177

55. **Pesticide Injury**
    David B. South 182

56. **Salinity Damage**
    Katy M. Mallams and Thomas D. Landis 185

Directory of Authors and Coordinators 189

Index of Nursery Pests 191

Index of Host Plants 194

Glossary 197

Diagnosis of Pest Problems
Alex C. Mangini

The Process

The nursery manager has a basic knowledge of symptoms and a habit of careful observation and can usually diagnose most forest nursery pest problems. The first step in diagnosing a pest problem is recognizing that a problem exists. To do this, the manager must routinely monitor the nursery and learn what healthy seedlings look like under all conditions. Once this healthy standard has been established, the manager (and properly trained personnel) can spot abnormalities. In addition, routine monitoring allows one to catch a problem early, when it is relatively simple and inexpensive to treat and before extensive damage occurs.

The next step is determining the cause. When a problem is found, the manager should use a systematic approach to evaluate probable causes. A systematic approach is more efficient than trying to select a single causal factor out of many possibilities. The key to this approach is to consider the most probable causes first; then, as each pest is ruled out, one moves logically to the less likely causes. This point is when knowledge of pest symptoms is crucial. Before any potential causes can be evaluated, however, it is necessary to do a thorough examination of both individual seedlings and the nursery in general.

Examination of the Nursery

The nursery manager should consider the location of the affected beds within the nursery as a whole. Are there any clues based on where the problem is located? For example, seedling damage in depressions may indicate problems such as poor drainage, pests that favor poor drainage conditions, or chemical buildup. The pattern of occurrence of symptoms and signs is important. Is the problem localized or widespread? For example, a pocket of affected seedlings with a progression of symptoms from the center outward may indicate a disease or insect infestation that is spreading. In contrast, seedling damage through the entire bed may indicate a physical problem such as nutrient deficiency, frost damage, or fertilizer burn. What cultural practices have been done in the affected area? Consider root pruning, fertilization, pesticide application, and other activities in this initial phase of the diagnostic process.

Examination of the Seedlings

Perhaps the most important part of the evaluation is examining the individual affected seedlings. Look carefully at foliage, stems, and roots of seedlings.

Start with the foliage; it is usually the first part of the seedling to show visible effects of an abnormality. Often, at this point, it is possible to determine if the problem is insect or disease related. Insect damage is usually obvious; either the actual insects or their conspicuous damage to the foliage are present. The foliage may be chewed, yellowed, or otherwise deformed by insect feeding; cast skins or frass may be present on the foliage.

Most foliage diseases caused by fungi, such as needle blight of conifers, and leaf spot, rust, and anthracnose on hardwoods, are characterized by small, discrete, usually darkened necrotic areas, often with fruiting bodies of the causal fungus developing on the dead tissue. Specialists use these fruiting bodies in identifying the causal agent and prescribing controls. Some fungi that cause needle spots on conifers, such as *Lophodermium* species, do not form fruiting bodies until most or all of the infected needle dies. The presence of needle spots on pines on which fruiting bodies never develop may indicate needle infection by stem rusts. Knowing the distribution and hosts of these fungi, which can be learned from information given in this manual, may be helpful in their diagnosis.

Other foliage symptoms, such as general chlorosis and needle tip necrosis on conifers, or large irregular necrotic blotches and marginal necrosis on hardwoods, are more difficult to diagnose. Air pollution, unfavorable environmental or soil factors, or root infection may cause these symptoms.

If a problem is found on the foliage, proceed to a close examination of seedling stems. Death or discoloration of the distal portion of the seedling or branch, with the remainder of the seedling appearing healthy, is a good indication of fungus, insect, or animal attack on the stem. Discrete necrotic areas characterize infection by pathogenic fungi, usually with a sharp line dividing the healthy and infected tissues. Fruiting bodies of the causal fungi are often present. Sun scald or chemical burns may mimic damage by pathogens, but the lesions of the former are often bleached rather than dark. Insect and animal attacks are usually readily identifiable by the feeding and gnawing wounds. Seedling tops and branch tips killed by pathogenic fungi are usually brick red, whereas those killed by insect or animal girdling are straw yellow.

Stem galls on pine seedlings are a good indication of stem rust infection. But galls on hardwood seedlings are likely to be caused by insects. If girdling by pathogenic fungi, insects, or animals occurs low enough on the stem, the entire top of the seedling will die. For this reason, it is important to examine seedlings as soon as they exhibit the first symptoms of chlorosis or wilting, when the root system may still be alive and the possibility of root problems can be more easily excluded.

Death of the entire top of the seedling usually indicates the presence of a root problem. Root-related problems are often the most difficult to diagnose because both abiotic and biotic factors can result in similar symptoms. For example, excessive soil moisture can result in symptoms similar to those from fungus or nematode attack. When seedlings exhibit symptoms, however, such as root blackening, lesions, shedding of the root cortex, or roughening of the bark on larger roots in the absence of obvious adverse soil factors, one would suspect fungus or nematode attack. Diagnosis of these problems usually requires laboratory analysis of plant tissue and soil because symptoms of most root diseases are similar and the causal agent is usually not evident from routine examination. The causal agent must be identified before proper control recommendations can be made.

Laboratory Analysis of Pest Samples

When the causal agent of seedling damage cannot be identified or when confirmation is desired, collect a representative sample and ship it to a plant diagnostic laboratory. Proper selection, handling, and packaging of samples to be submitted are crucial steps in a correct and timely diagnosis. The following guidelines will be useful in this regard:

1. Select 10 apparently healthy, 10 moderately affected, and 10 severely affected seedlings. Wrap each group of 10 seedlings in moist paper towels or similar wrapping material, label, and place them in a plastic bag.

2. Place a sample of most insects observed on the affected seedlings in vials of alcohol and include them with the seedlings. Mites, scales, aphids, and caterpillars should be sent in alive on some of the infested foliage or stems placed in a plastic bag.

3. Collect soil samples from the affected area and from unaffected portions of the beds. Separate soil samples should be obtained from the seedling root zone. Do not include the top crust of soil.

4. Ship the samples in a durable cardboard box or similar container by the fastest means available. Seedling samples should be shipped with refrigerant to prevent overheating and development of mold.

If possible, include the following information with each sample:

1. Species.
2. Age.
3. Present nursery production quantity (thousand seedlings).
4. Percent of seedlings affected.
5. Date the symptoms were first observed.
6. Names of other affected species.
7. Pesticides, with dosage rates and application dates.
8. Fertilizers, with dosage rates and application dates.
9. Possible weather problems.
10. Cultural practices recently used, such as root pruning.
11. Fumigation history (fumigant, formulation, dosage, rates, dates, and season).
12. Soil type.
13. Soil analysis results (concentrations of macronutrients, micronutrients, organic matter, and pH).
14. Signs (fungi or insects) and disease symptoms noted on the foliage, stems, and roots.
15. General development of ectomycorrhizal feeder roots.

Laboratory diagnosis of samples is usually rapid but may sometimes take several weeks. This method is a multistage diagnostic process, with time lapses between stages to allow the fungal pathogen or insect to develop under controlled laboratory conditions.

Pest Selection Key

The information included in the Diagnosis of each chapter in this manual is intended to guide the nursery manager or pest specialist to the cause of an observed problem. As mentioned previously, isolating the cause can be greatly facilitated by a systematic assessment of the symptoms and signs and other information available for a specific problem. The following Pest Selection Key (page 4) will help.

Three kinds of information are needed to use the key:

1. The part of the plant affected—seed or cone, roots, stem, and foliage.

2. The symptoms and signs observed—chlorosis, dead tops, leaf spots, swelling, or galls, etc.

3. The type of seedling on which the problem occurs—conifer or hardwood.

The numbers following the host designation refer to the chapter in the text where pests that cause the type of damage described in the key can be found. Some pests may affect more than one part of the plant, cause more than one kind of symptom, or occur on both hardwood and conifer seedlings. Cross-referencing the symptoms can eliminate some pests or problems. For example, for information about a conifer seedling with yellowing foliage (chapters 2, 3, 7, 11, 13, 14, 18, 19, 32, 34, 36, 37, 38, 41, 42, 45, 50, 51, 52, 54, and 56) and tip dieback (chapters 3, 10, 11, 12, 14, 17, 21, 32, 34, and 40), one would consult only the chapters represented in both groups of symptoms—3, 11, 14, 32, and 34.

Selected References

Anderson, R.L.; Cordell, C.E.; Landis, T.D.; Smith, Jr., R.S. 1989. Diagnosis of pest problems. In: Cordell, C.E.; Anderson, R.L.; Hoffard, W.H.; Landis, T.D.; Smith, Jr., R.S.; Toko, H.V.; tech. coords. Forest nursery pests. Agriculture Handbook 680. Washington, DC: USDA Forest Service: 1–4.

Diagnosis of Pest Problems

Pest Selection Key

Damage Categories			Seedling Type	
			Conifer	Hardwood
			Chapters	
Foliage	Normal color	Partly missing	21, 22, 43, 51	30, 43, 51
		Insect feeding	22, 43, 51	29, 30, 43, 51
		Insects present	21, 22, 42, 46	29, 30, 42, 46
	Foliage discolored	Stunted	2, 31, 32, 36, 37, 38, 48, 51, 54, 55, 56	31, 32, 36, 37, 38, 51, 54, 55, 56
		Spots present	1, 4, 7, 8, 10, 12, 15, 19, 50	23, 24, 25, 26, 27, 32, 40, 41, 50
		Defoliation	1, 4, 7, 8, 10, 32, 50, 54	23, 24, 25, 26, 27, 28, 41, 50, 54
		Partly dead	1, 2, 3, 4, 7, 8, 9, 10, 13, 15, 16, 18, 21, 22, 35, 55, 56	23, 24, 25, 30, 35, 40, 41, 55, 56
		All dead	1, 2, 3, 8, 11, 15, 18, 31, 32, 35, 37, 38, 55	23, 31, 32, 33, 35, 37, 38, 40, 41, 55
		Deformed	22, 42, 48, 55	23, 24, 28, 41, 42, 55
		Red/brown	2, 3, 7, 8, 9, 10, 12, 13, 14, 15, 16, 17, 18, 19, 21, 22, 31, 32, 35, 37, 38, 50, 54, 55	31, 33, 35, 37, 38, 40, 50, 54, 55
		Yellow	2, 3, 7, 11, 13, 14, 18, 19, 32, 34, 36, 37, 38, 41, 42, 45, 50, 51, 52, 54, 56	28, 32, 34, 36, 37, 38, 40, 41, 42, 45, 50, 51, 52, 54, 56
		Wilted	14, 31, 33, 34, 37, 38, 40, 45, 47, 52, 54	31, 33, 34, 37, 38, 40, 45, 47, 52, 54
		Cottony or silky material present	15, 18, 21, 35, 50	28, 35, 50
		Fungal fruiting bodies visible	1, 3, 4, 5, 8, 10, 12, 16, 17, 32, 35	23, 26, 27, 28, 32, 35
		Insects present	42, 50	42, 50
		Insect frass present	21	29, 30
Stem	Swelling (galls)		5, 6, 42, 46, 54, 55	42, 46, 48, 50, 54, 55
	Bark missing		20, 35, 44, 46, 47, 51, 53	35, 44, 46, 47, 51, 53
	Stem cut or broken		43, 44, 47, 53	29, 33, 43, 44, 47, 48, 53
	Sunken areas and/or discoloration		3, 6, 11, 13, 14, 16, 17, 19, 32, 33, 34, 35, 43, 54	23, 24, 25, 26, 32, 33, 34, 35, 37, 40, 54
	Tip dieback		3, 10, 11, 12, 14, 17, 21, 32, 34, 40	23, 24, 26, 30, 32, 34, 40
	Insects present in or on stem		46	29, 46
	Stem deformed		3, 12, 13, 17, 33, 38, 48	29, 38, 48
	Fungal fruiting bodies visible		13, 32, 34	32, 34
Root	Root collar dead		3, 31, 37	31, 37
	Primary roots dead or missing		2, 32, 34, 37, 38, 47, 52, 53, 54	32, 34, 37, 38, 47, 52, 53, 54
	Fine roots dead or discolored		2, 31, 32, 34, 37, 38	31, 32, 34, 37, 38
	Fine roots missing		2, 31, 34, 37, 38, 45, 51, 52	31, 34, 37, 38, 45, 47, 51, 52
	Roots swollen		31, 36	29, 31, 36
	Roots stripped		20, 32, 34, 38, 45, 52, 54	32, 34, 38, 45, 52, 54
	Insects In or on roots		45, 51	29, 45, 51
	Fungal fruiting bodies visible		31	31
Seeds or cones			33, 34, 39, 44, 49, 53, 55	33, 34, 39, 44, 49, 53, 55
Soil	Disturbed		44, 47, 53	44, 47, 53
	Discolored		56	56

Integrated Nursery Pest Management

R. Kasten Dumroese

What is integrated pest management? Take a look at the definition of each word to better understand the concept. Two of the words (integrated and management) are relatively straightforward. Integrated means to blend pieces or concepts into a unified whole, and management is the wise use of techniques to successfully accomplish a desired outcome. A pest is any biotic (biological) stress factor that interferes with healthy seedling development and causes a sustained departure from the normal physiological or morphological condition that characterizes a healthy seedling. So pests can be microorganisms (for example, fungi, bacteria, viruses), weeds, and animals (for example, insects, nematodes, rodents, deer, and a well-meaning, but inept-tractor operator or chemigator). Therefore, integrated pest management (IPM) incorporates a variety of techniques to limit losses caused by a variety of biotic stresses to successfully produce a healthy seedling crop.

Most nurseries have the same comprehensive goals of producing high-quality seedlings in a cost-effective manner, and ensuring employee and environmental safety. Formalizing those goals ensures that any pest treatments are consistent with those objectives. Prudent nursery managers also have a formal, clearly articulated, decisionmaking or management policy. Such a document transcends changes in staffing, and when combined with historical records (see the next section), can assist nursery managers in making correct pest management decisions. Nursery pests are always present; the best IPM plans describe potential pests and define the critical threshold for the pest to be classified as a problem. After a pest population crosses that designated threshold, a series of control options can be applied.

Training, Monitoring, and Recordkeeping

Employees should be encouraged to attend extension workshops and regional nursery meetings. These sessions offer opportunities to hear the latest about new pest problems, control techniques, pesticides, and other important nursery topics, such as fertilization and irrigation. Such gatherings also provide excellent opportunity for employees to network among other nursery professionals, and often the information gleaned during conversations held during breaks at these meetings is invaluable. Generally, such gatherings are also more focused on pests encountered in forest and conservation nurseries, as compared with generic "pest" workshops required to maintain pesticide applicators licenses. Prudent managers can combine IPM training and implementation as part of their overall U.S. Environmental Protection Agency Worker Protection Standard training. Using a variety of IPM techniques to reduce the need for chemical applications (that require reentry intervals) has advantages; supervisors will spend less time notifying employees about upcoming restrictions and access to the entire facility by workers can be better maintained.

Early detection of pest problems often results in pest control before damage is severe. All employees are responsible for continually monitoring the seedlings for pests and factors that disrupt appropriate environmental conditions for healthy seedling growth. After pests are detected, designated employees who are specifically trained to identify and document problems can recommend appropriate IPM treatments. These employees need to be equipped with a hand lens; small notebook, personal digital assistant, or digital recorder; digital camera; and maybe even a global positioning system receiver in larger nurseries (fig. I.1). If nursery staff cannot make an accurate diagnosis, samples should be sent immediately to appropriate pest specialists for identification. Proper pest identification is essential in selecting the most cost-effective, practical, economical, and environmentally sound IPM practice. Full details about the pest incidence should

Figure I.1—*Observant employees and a hand lens are critical to any integrated pest management plan.* Photo by Niklaas K. Dumroese.

be recorded and retained in a central location. Historical records are valuable assets used to accurately diagnose future problems and increase the effectiveness of subsequent treatments.

Managing Pests

Because good IPM plans incorporate many pest control techniques, this holistic approach is more practical, economical, and ecologically sound than a control program that relies, for example, on a single control technique. IPM can be discussed and implemented many ways. An effective IPM plan focuses on the three elements necessary for pests to cause problems in nurseries: (1) the pest is present, (2) the environment is conducive for the pest, and (3) the hosts (nursery crops) are susceptible to the pest (fig. I.2). These collective elements are referred to as the "disease triangle" because all three elements are necessary for biotic disease to occur. A good IPM program, therefore, coordinates and targets multiple control methods against all three components. Moreover, using multiple methods to control pests makes the nursery crops less vulnerable and more resilient when either litigation or regulation remove or restrict the ability to use a particular pest prevention technique.

It is impossible to provide an IPM "recipe" for all nurseries, simply because each nursery has unique goals and different ideas of what constitutes acceptable pest populations or damage thresholds. Additionally, each nursery has varying pest issues based on its geographic location and associated environment. Some general information is provided in the following sections to help nursery managers and their workers gain an improved understanding of the wide scope of activities that can be part of an IPM plan. Keep in mind that this discussion is far from an exhaustive list of potential techniques; hopefully, the examples provided will stimulate additional, innovative techniques. Some techniques described may block more than one component of the disease triangle. Because separating the interaction of seedling health and proper environmental conditions is often difficult, those two components of the disease triangle are combined in the following discussion; these two components also are the logical starting point for any IPM plan because they primarily focus on pest prevention rather than what can be done after a pest is causing damage.

Providing a Healthy Environment and Improving Seedling Resistance and Resilience

A healthy environment and seedling resistance and resilience are intertwined. When seedling damage is noted but pests are absent, the cause can often be traced to an environmental factor, which highlights the importance of examining these two components concurrently.

Abiotic (environmental) and biotic (living) factors contribute to seedling health. For a bareroot nursery or a container nursery without greenhouses, site selection is the most important abiotic variable to consider. Having a site that meets criteria for optimum soil texture, pH, drainage, air flow, and water quality (low salinity, proper pH) contributes to healthy seedlings and is paramount in selecting new sites or expanding existing sites. Optimum soil is well drained, low in clay content, and lacks any impermeable layers (especially close to the

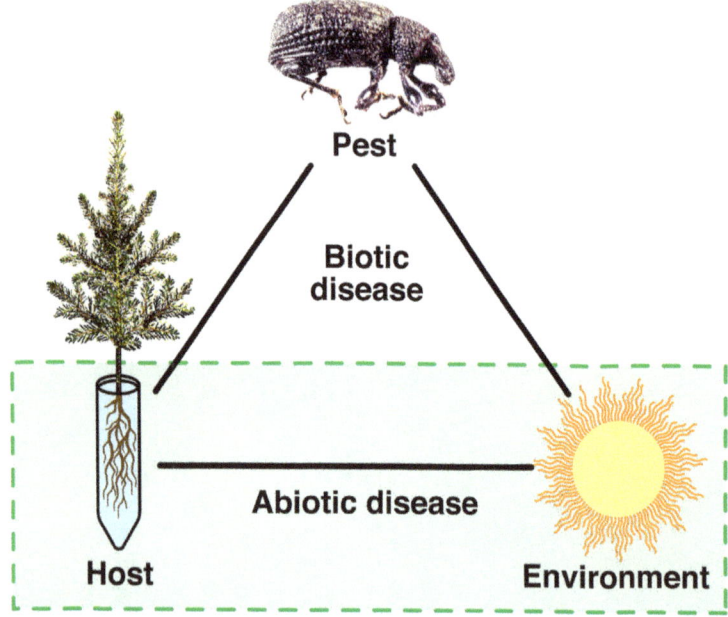

Figure I.2—The "disease triangle" illustrates the concept that a host, a pest, and a conducive environment are necessary to cause biotic disease. Abiotic disease occurs when environmental factors, such as frost, injure the host plants. Illustration by Jim Marin Graphics.

surface). Poorly drained soils can lead to root rot development in bareroot nurseries or can result in slow drainage from a container nursery site. If poor drainage is a problem in a portion of the nursery, subsoiling (deep plowing to fracture impermeable layers) or tiling may mitigate the problem. Fields may also be leveled or seed beds oriented to improve water flow across the site.

Although soil pH can be modified, beginning with a pH conducive for seedlings (generally 4.5 to 6.5 depending on species) is best. For conifers, a pH above 5.5 generally leads to more damping-off and root rot, and a pH above 6 can reduce ectomycorrhizae development. A good site is not in a frost pocket and is protected from harsh winds.

For any nursery type, water needs to be tested for pH and the presence of detrimental fungi, nematodes, insects, weed seeds, salts, and pesticides, especially if a surface source (river, pond, or reservoir) is used (fig. I.3). Although pH can be readily adjusted downward by injecting acid into irrigation water, it may be difficult, expensive, or impossible to remove these other contaminants; test results can help managers use appropriate methods to mitigate their influence.

Proper temperatures are required for optimum seedling growth. Improper temperatures can reduce germination, induce fungal attack, or directly cause seedling mortality. For crops grown in greenhouses equipped with environmental controls, temperature is easily maintained by a combination of heaters, air conditioners, cooling pads, removable roofs, shading, and irrigation. For bareroot nurseries and outdoor container operations, maintaining desired temperatures is difficult. High temperatures are usually mitigated by irrigation, although some species may require shading provided by laths. Low temperatures in early spring can be avoided by delaying sowing until the soil temperature is suitable for rapid seed germination and early seedling growth. Damage from an early frost in the fall can be mitigated by applying irrigation water. Using a cover crop can also help protect germinating seeds from cold temperatures; wheat or rye, planted with fall-sown crops, particularly hardwoods, germinates in the fall and provides a thermal blanket for the crop seeds. Similarly, insulating fabrics can protect fall- or spring-sown crops from excessively cold temperatures; in greenhouses, such fabric can help retain nighttime media temperatures and foster more rapid germination with less energy consumption.

Strong, healthy plants are more resilient to pests. Choosing species and seed sources appropriate to the nursery location is important (especially for bareroot nurseries); seedlings grown off site will be more susceptible to environmental stresses. Vigorous seed lots that have rapid and uniformly high germination are often more immune to biotic and abiotic stresses. Seeds produced in seed orchards usually have high vigor, and often the clones have been selected for resistance to pests, such as fusiform rust.

Crop density should be controlled to ensure seedlings have access to sufficient resources (nutrients, water, and sunlight), whether they are grown in bareroot or container nurseries. High crop density can reduce seedling vigor, making them more susceptible to pests. In addition, intermingling of canopies can promote high relative humidity, which is conducive to foliar diseases, such as powdery mildew and gray mold.

Water is the most dangerous input used in nurseries. Insufficient irrigation can lead to improper seedling nutrition, poor growth, greater susceptibility to pests, and mortality. Excessive irrigation, which reduces gas exchange to and from soil and media, can kill roots and

Figure I.3—*Nursery water needs to be tested for pH, salinity, pesticides, and the presence of fungi, nematodes, insects, and weed seeds, especially if a river, pond, or reservoir is the water source.* Photo by Thomas D. Landis, USDA Forest Service.

reduce plant vigor, enabling waterborne organisms (for example, *Pythium*, *Phytophthora*, and some nematodes) to become pests. *Pythium* spores require water to move to new hosts, so saturated soils are especially conducive to spread of this disease. Excessive irrigation also favors shoot development at the expense of root development, which could decrease mycorrhizae development. Irrigation timing can also be critical. Applications made early in the day enable the foliage to dry, which decreases the likelihood of foliar diseases, such as gray mold.

Providing crops with proper nutrition levels is paramount. In addition to knowing proper nutrition levels and which fertilizer formulations work best for specific species, employees need training to efficiently and accurately administer applications. Insufficient fertilizer use can reduce seedling vigor and growth and increase susceptibility to pests. Excessive fertilizer use can stimulate damping-off during germination, burn foliage, stunt growth, cause excessive shoot growth and diminish root growth or, in the case of nitrogen and phosphorus, delay or inhibit formation of symbiotic relationships, such as mycorrhizae. Optimum rates yield the healthiest seedlings.

Symbiotic microorganisms (for example, mycorrhizae and nitrogen-fixing bacteria) can improve nutrient and water uptake toward enhancing seedling health. Colonization by symbiotic microorganisms and other favorable biotic species (for example, *Trichoderma* to reduce root disease incidence) can reduce potential infection by pathogenic fungi.

Managers should properly harden crops prior to harvest to reduce seedling stress levels. Keeping seedlings cool and moist and handling them with care during harvest ensures seedling quality is retained. Seedlings are more susceptible to stresses and pests when roots are allowed to dry or seedling boxes are dropped onto the ground.

Nursery managers often realize that portions of their container or bareroot nurseries are more favorable to growing some species than other areas. Growing seedlings in areas where the environmental and biotic features are more conducive to that species yields healthier, more pest-resistant crops. Continuous cropping of a particular species in the same area, especially in bareroot nurseries, can promote the buildup of a variety of fungal, insect, nematode, and weed pests. Rotating crops, especially susceptible and nonsusceptible species, and alternating seedling production with cover crops is one way to prevent such buildup.

Adding organic mulches is one method for improving soil organic matter levels; improved nursery soil tilth enhances aeration, water-holding capacity, and improved soil microorganism populations that may help suppress root disease. Commonly used mulches include pine needles, compost, sawdust, and straw. Nursery staff should have any mulch material tested to ensure it is free of pests, especially fungal inoculum and weed seeds. Although mulches can be fumigated, it is probably more efficient to incorporate any organic material before the traditional presowing fumigation occurs. Avoid mulches with high carbon-to-nitrogen ratios (for example, fresh sawdust or straw) that can lead to nitrogen immobilization as microorganisms use the nitrogen to digest the mulch.

Reducing Pest Populations

Pests range in size from microscopic (fungi, nematodes, bacteria, and viruses) to small (insects) to megafauna (rabbits and deer), and methods for reducing pest populations are just as varied. Often, nursery managers equate pest control or pest management with a repeated pesticide application regime. At some nurseries, such applications are made whether the pest is present or not; some managers consider it to be cheap insurance. Other nurseries have discovered, however, that reducing initial pest populations through other techniques and resorting to chemicals only when pests cross a critical damage threshold, can dramatically reduce the amount of pesticides required. This approach can result in financial savings and less down time because of required reentry intervals after pesticide applications.

Physical: Sanitation and Exclusion

One easy method to reduce pests is by increasing sanitation across the nursery. Sanitation can take many forms, including keeping the landscape, seed production areas, and nursery perimeter free of weeds, which can reinfest bareroot and container crops and harbor other insect and fungal pests. Maintaining growing areas and nearby grounds free of accumulated junk reduces hiding places for larger pests, such as rodents, that can consume seeds or seedlings. Clean field machinery before moving from one field to another to help prevent movement of weeds, nematodes, and fungal pathogens. Employees can rogue dead and dying seedlings from beds and containers to reduce the spread of disease. Diseased culls should be buried or composted at a spot away from the growing area.

Employees should wash their hands to prevent spreading contaminants when moving from one seed source to another. In container nurseries, tables, floors, and the containers need to be cleaned between crops. Solid floors are easier to keep clean and prevent weeds and some insects, such as fungus gnats that commonly reproduce in the organic matter under benches. Fungus gnats can consume seedlings roots and act as vectors for Fusarium root disease and gray mold by carrying spores on their bodies. Reused containers that have not been cleaned often reduce seedling growth, even if traditional root disease symptoms are absent.

Some pests are easier to exclude than others. Screens over entry points into greenhouses can exclude many insects, and fences can be used to prevent larger animals from damaging plants. Seed germination cloths or lath or cage structures deployed over seedbeds and containers in outdoor growing areas can reduce bird predation until germination is complete (fig. I.4). Refraining from accepting stock from other nurseries can also reduce the likelihood of introducing pests. Some pests can also be excluded through trapping. Aphids, fungus gnats, thrips, whiteflies, and others can be lured to colored sticky traps; these traps can also serve to monitor population levels and control these pests. Rodents can be removed using lethal or humane traps. Nursery managers in the Midwest use thick mulches of straw above their fall-sown bareroot crops of oaks and hickories to exclude turkey and deer predation.

Although windbreaks around nurseries can improve aesthetics, reduce soil erosion, and decrease wind damage, careful evaluation of the windbreak species is important to ensure that candidates are neither an alternate host for fungi nor a food source or refuge for other pests. For example, oaks serve as alternate hosts for fusiform rust, which can infect southern pine crops. Austrian pine, commonly planted as a windbreak tree in the Lake States, is susceptible to Diplodia blight, which can severely damage pine seedlings. Therefore, judicious windbreak species selection is necessary, as is keeping the windbreak healthy and vigorous so that pests are less likely to become a problem. Unfortunately, spores of these fungal pests can travel long distances, making complete eradication, even to extended distances around the nursery, impractical to achieve.

Weeds are often excluded from fallow nursery fields by repeated tillage or through suppression using a cover crop. Cover crops reduce erosion and increase soil organic matter but need to be selected carefully. Intuitively, legumes that fix nitrogen seem to be good candidates for cover crops, but they are also known to promote development of charcoal root rot, black root rot, and nematodes in the Southern States, and Fusarium root disease in the Pacific Northwest. This concern may be irrelevant if fields with legume cover crops are plowed under and enabled to decompose, or if the fields are fumigated before establishing the next seedling crop.

Biological Control

Biological controls can also reduce pest populations. Incidence of root rots can be reduced by drenching soil or media with *Trichoderma*, a species antagonistic

Figure I.4—*A screen exclosure prevents seed predation at the Hawai'i Division of Forestry and Wildlife Kamuela (Waimea) State Tree Nursery on the Island of Hawai'i.* Photo by R. Kasten Dumroese, USDA Forest Service.

to many pathogenic fungi. Inoculating crops with mycorrhizal fungi may help reduce incidence of root rots, either by improving host resistance or by decreasing the number of potential infection points on seedling roots. Biological control of insects is a bit more difficult. Although many insect pests can be controlled with beneficial insects, ensuring that sufficient beneficial insects are in place at the correct time to counteract undesirable insects can be difficult to achieve. Often, a minimal pest level is required to ensure that the biological agent survives, and this minimal level may exceed acceptable damage thresholds. Nursery managers must exercise care to ensure that other pest control measures do not diminish the efficacy of potential biological controls. For example, some systemic fungicides can suppress mycorrhizae development. Predator nematodes are available for biological control of fungus gnats in container nurseries and harmful nematodes in fields.

Chemicals

Chemical treatments are another piece of any IPM plan and are useful in situations in which other IPM practices are unavailable or ineffective (either from a pest control or financial aspect). Chemical pesticides (biocides, herbicides, insecticides, rodenticides, and nematocides) can be synthetic (for example, glyphosate) or natural or botanic (for example, neem oil). This distinction is becoming blurred, as many botanic chemicals are now produced synthetically, rather than extracted from their natural sources.

Chemicals can be applied to seeds to protect them against birds, animals, and fungi. Thiram is one such chemical; it decreases predation by birds and small animals and has efficacy against some seed- and soil-borne pathogenic fungi. Many other chemicals (for example, bleach, hydrogen peroxide, and fungicides) have been used on forest tree seeds; their efficacy varies by tree species, pathogen, chemical concentration, and treatment duration. In the South, systemic fungicides are commonly used to treat loblolly and slash pine seeds so that newly emerged seedlings are protected from fusiform rust.

Chemicals can be applied to soil several ways. Fumigants are nonselective biocides. Fumigating soil before sowing effectively reduces or eliminates fungi, insects, nematodes, and weed seeds. Although fumigants are effective in reducing or eliminating soilborne pathogenic fungi and weed seeds, they also negatively affect beneficial microorganisms, such as mycorrhizal fungi and *Trichoderma* species. Usually fields are rapidly recolonized by ectomycorrhizal fungi because the fungi reproduce through windborne spores, therefore, most conifer crops become mycorrhizal. Spores of arbuscular mycorrhizal fungi are not airborne, so they take much longer to recolonize hardwood (and some conifers, such as giant sequoia and bald cypress) nursery beds. Arbuscular mycorrhizal fungi can recolonize nursery beds from adjacent, nonfumigated soils, but applying arbuscular mycorrhizal inoculum after fumigation can ensure seedlings have their symbiotic fungi.

Pre- and post-emergent herbicides are commonly applied to the soil immediately before and after germination of crop seeds to prevent weed seed germination. Judicious, conscientious, and safe herbicide use has reduced the need for more expensive hand weeding. Some pre- and post-emergent herbicides can be used with a variety of hardwood and conifer crops in bareroot and container nurseries. Nonfumigant fungicides are less commonly applied to soil, and insecticides are rarely applied, if ever. When used, these pesticides may be applied as soil drenches or granular formulations. Fungicides are sometimes applied in bareroot and container nurseries to control root rot, but their efficacy is often low or poor, mostly because it is difficult to move these short-lived products to sufficient soil depths in sufficient quantities to be useful.

Chemicals can be applied to plants either to protect crop plants (fungicides and insecticides) or to kill weeds (herbicides). Fungicides and insecticides can be contact (kills the pest on contact) or systemic (absorbed by the seedling, and the pest is killed when it consumes a portion of the crop plant) pesticides. Typically fungicides and insecticides are applied with spray equipment although sometimes they are injected into irrigation lines. Contact pesticides have low residual activity and therefore usually require multiple applications, especially to protect new growth or if reinfestation by the pest occurs. Systemic pesticides, translocated throughout a seedling, have longer residual activity. Although simple hydraulic sprayers are commonly used to apply chemicals, electrostatic and other specialty sprayers can reduce water and pesticide volumes and increase pesticide efficacy (fig. I.5). Herbicides can reduce weeds in containers, fields, and other nursery areas. Nonselective herbicides are often used in places where crops are not involved, such as along irrigation lines, fences, and near buildings, but selective herbicides are used with seedling crops. Recent development of "shielded sprayers" enables nursery personnel to apply nonselective herbicides to weeds between conifer or hardwood rows in nursery beds.

Figure I.5—*This modified orchard sprayer applies pesticides thoroughly to seedlings being grown at the Missouri Department of Conservation George O. White Nursery at Licking.* Photo by R. Kasten Dumroese, USDA Forest Service.

Embrace Integrated Pest Management

Many good nursery managers practice IPM without realizing it or calling it by name. IPM becomes easier to apply if nursery managers make an effort with themselves and their employees to embrace IPM as a management philosophy—a priority. Doing so will enable nursery managers to reap the maximum benefit of any IPM program.

Selected References

Barnett, J.P. 2008. Relating seed treatments to nursery performance: experience with southern pines. In: Dumroese, R.K.; Riley, L.E.; tech. coords. National proceedings: forest and conservation nursery associations—2007. RMRS-P-57. Fort Collins, CO: USDA Forest Service, Rocky Mountain Research Station: 27–37.

Cordell, C.E. 1979. Integrated control procedures for nursery pest management. In: Proceedings, Northeastern Area nurserymen's conference. Broomall, PA: USDA Forest Service, Northeastern Area, State and Private Forestry: 43–51.

Cordell, C.E.; Filer, Jr., T.H. 1985. Integrated nursery pest management. In: Lantz, C.W., ed. Southern pine nursery handbook. Atlanta: USDA Forest Service, State and Private Forestry, Region 8. Chapter 13.

Cordell, C.E.; Kelley, W.D.; Smith, Jr., R.S. 1989. Integrated nursery pest management. In: Cordell, C.E.; Anderson, R.L.; Hoffard, W.H.; Landis, T.D.; Smith, Jr., R.S.; Toko, H.V., tech. coords. Forest nursery pests. Agriculture Handbook 680. Washington, DC: USDA Forest Service: 5–13.

Cram, M.M.; Fraedrich, S.W. 2005. Management options for control of a stunt and needle nematode in Southern forest nurseries. In: Dumroese, R.K.; Riley, L.E.; Landis, T.D., tech coords. National proceedings: forest and conservation nursery associations—2004. RMRS-P-35. Fort Collins, CO: USDA Forest Service, Rocky Mountain Research Station: 46–50.

Dumroese, R.K. 2008. Observations on root disease of container whitebark pine seedlings treated with biological controls. Native Plants Journal. 9: 92–97.

Dumroese, R.K.; James, R.L. 2005. Root diseases in bareroot and container nurseries of the Pacific Northwest: epidemiology, management, and effects on outplanting performance. New Forests. 30: 185–202.

Dumroese, R.K.; James, R.L.; Wenny, D.L. 2002. Hot water and copper coatings in reused containers decrease inoculum of *Fusarium* and *Cylindrocarpon* and increase Douglas Fir seedling growth. Horticultural Science. 37: 943–947.

Dumroese, R.K.; Pinto, J.R.; Jacobs, D.F.; Davis, A.S.; Horiuchi, B. 2006. Subirrigation reduces water use, nitrogen loss, and moss growth in a container nursery. Native Plants Journal. 7: 253–261.

Dumroese, R.K.; Wenny, D.L.; Quick, K.E. 1990. Reducing pesticide use without reducing yield. Tree Planters' Notes. 41(4): 28–32.

Enebak, S.; Carey, B. 2002. Pest control for container-grown longleaf pine. In: Barnett, J.P.; Dumroese, R.K.; Moorhead, D.J., eds. Proceedings of workshops on growing longleaf pine in containers—1999 and 2001. GTR-SRS-56. Asheville, NC: USDA Forest Service, Southern Research Station: 43–46.

Hansen, E.M.; Hamm, P.B.; Campbell, S.J. 1990. Principles of integrated pest management. In: Hamm, P.B.; Campbell, S.J.; Hansen, E.M., eds. Growing healthy seedlings: identification and management of pests in northwest forest nurseries. Corvallis, OR: Oregon State University, Forest Research Laboratory. Special Publication 19: 92–105. Chapter 33.

Hawkins, B. 2004. Use of living mulches to protect fall-sown crops. Native Plants Journal. 5: 171–172.

James, R.L.; Dumroese, R.K.; Wenny, D.L. 1990. Approaches to integrated pest management of Fusarium root disease in container-grown seedlings. In: Rose, R.; Campbell, S.J.; Landis, T.D., eds. Target seedling symposium: proceedings, combined meeting of the Western Forest Nursery Association and Intermountain Nursery Association. GTR-RM-200. Fort Collins, CO: USDA Forest Service, Rocky Mountain Forest and Range Experiment Station: 240–246.

James, R.L.; Dumroese, R.K.; Wenny, D.L. 1993. Principles and potential for biocontrol of diseases in forest and conservation nurseries. In: Landis, T.D., tech. coord. Proceedings, Western Forest Nursery Association. GTR-RM-221. Fort Collins, CO: USDA Forest Service, Rocky Mountain Forest and Range Experiment Station: 122–131.

James, R.L.; Dumroese, R.K.; Wenny, D.L. 1995. *Botrytis cinerea* carried by adult fungus gnats (Diptera: Sciaridae) in container nurseries. Tree Planters' Notes. 46(2): 48–53.

Kelley, W.D.; Cordell, C.E. 1984. Disease management in forest tree nurseries. In: Proceedings, integrated forest pest management symposium. Athens, GA: University of Georgia, Center for Continuing Education: 238–246.

Landis, T.D.; Luna, T.; Dumroese, R.K. 2008. Holistic pest management. In: Dumroese, R.K.; Luna, T.; Landis, T.D., eds. Nursery manual for native plants: a guide for tribal nurseries. Vol. 1: nursery management. Agriculture Handbook 730. Washington, DC: USDA Forest Service: 262–275.

Landis, T.D.; Tinus, R.W.; McDonald, S.E.; Barnett, J.P. 1990. The biological component: nursery pests and mycorrhizae, vol. 5. The container tree nursery manual. Agriculture Handbook 674. Washington, DC: USDA Forest Service. 171 p.

Schmal, J.L.; Woolery, P.O.; Sloan, J.P.; Fleege, C.D. 2007. Using germination cloths in container and bareroot nurseries. Native Plants Journal. 8: 282–286.

South, D.B. 1984. Integrated pest management in nurseries—vegetation. In: Proceedings, integrated forest pest management symposium. Athens, GA: University of Georgia, Center for Continuing Education: 247–265.

South, D.B. 2009. A century of progress in weed control in hardwood seedbeds. In: Dumroese, R.K.; Riley, L.E., tech. coords. National proceedings: forest and conservation nursery associations—2008. RMRS-P-58. Fort Collins, CO: USDA Forest Service, Rocky Mountain Research Station: 80–84.

South, D.B.; Enebak, S.A. 2006. Integrated pest management practices in southern pine nurseries. New Forests. 31:253–271.

Sutherland, J.R. 1984. Pest management in Northwest bareroot nurseries. In: Duryea, M L.; Landis, T.D., eds. Forest nursery manual: production of bareroot seedlings. Boston, MA; The Hague, The Netherlands; Lancaster, PA: Martinus Nijhoff/Dr. W. Junk Publishers: 203–210. Chapter 19.

Wichman, J.; Hawkins, R.; Pijut, P.M. 2005. Straw mulch prevents loss of fall-sown seeds to cold temperatures and wildlife predation. Native Plants Journal. 6: 282–285.

Evaluating Damage to Nursery Crops

Katy M. Mallams and Tom Starkey

In spite of the best efforts at prevention and control, all nurseries experience crop damage due to pests. But the question arises: does the damage in a particular nursery justify additional expenditures for prevention or control? To answer that question, managers must identify the damage and evaluate costs and benefits of prevention and treatment and impacts on outplanting success. Production of seedlings in the nursery is an integral part of forest regeneration and restoration programs. Damage to seedlings in the nursery may result in crop losses or poor field performance after outplanting in the field. Poor performance after outplanting may also be caused by damage occurring after seedlings leave the nursery. Diligent recordkeeping and evaluation and reporting of seedling performance is critical to correctly identify the cause of damage and when, where, and why it occurred, so the most effective and least costly treatments can be applied and future damage prevented.

Evaluating Damage in the Nursery

Damage caused by pests in forest nurseries can affect crops in several ways. Outright seedling mortality is the most obvious impact. Physical damage and growth loss may be less obvious, but are no less important. Seedlings that have been killed or are otherwise unsuitable for planting are usually a total loss. Occasionally, seedlings that are too small or have minor physical damage can be transplanted. Other seedlings may meet specifications and be shippable in spite of damage. The objective of evaluating crops for damage is to identify whether seedlings are suitable for outplanting or transplanting or should be culled. The information can also be used to make decisions about the efficacy of control treatments in the current crop and to plan preventive actions in future crops. To determine the impact of damage, and appropriate treatment and prevention methods, correct diagnosis of causal agents is critical.

A program for identifying and evaluating pest damage should be a routine part of nursery operations. Such evaluations may be designed either for general pest occurrence or for specific problems. The intensity of an evaluation may range from informal seedbed scouting to intensive sampling procedures that permit statistical analyses. Preliminary sampling data may be needed to determine the distribution and variation of the damage to select the most effective and efficient evaluation. Causal agents may include biotic factors such as insects, pathogens, and animals, and abiotic factors such as the nursery environment, mechanical damage, and chemical agents. A person knowledgeable in identification of damaging agents should supervise or conduct the evaluations. It is important to conduct evaluations when symptoms of damage are readily visible. This step may require surveys several times during the crop production cycle. Proper timing of evaluations requires understanding of local nursery insect and disease problems. Insects that cause damage, such as lygus bugs, often appear in early summer. Root diseases, stem rusts, and foliage blights are easiest to see late in the growing season. Fungal diseases, especially root diseases, can manifest subtly with chlorosis or stunting. Fungal diseases may be difficult to diagnose in the field and may require whole plant testing in a laboratory. Some belowground damage, including that caused by root diseases, root weevils, and cranberry girdlers, may not be visible until seedlings are lifted or packed.

Keeping a thorough and accurate record of evaluations is very important. Information to record includes the date, seedbed location, seedling species, seedlot, family, stock type, signs and symptoms observed, causal agent, age of the affected seedlings, portion of seedlings affected, and the size and distribution of the affected area. Background information that is helpful both for planning evaluations and diagnosing damage includes soil type and drainage, irrigation records, cultural practices, pesticide and fertilizer use, and weather data.

After a diagnosis is made, consideration of costs and benefits and analysis of environmental impacts is needed to select treatments that are the most economical and safest for nursery field workers and the environment. The loss in crop value resulting from seedling damage and mortality is affected by the extent of the damage, the value of the seed and seedlings, and the costs associated with transplanting and treatments to control pests. Seed value, seedling age, and stock type are significant factors because they represent the level of investment that has already been made in a crop. The cost of control treatments and transplanting represent additional investments that may be needed before a damaged crop can be outplanted.

Use of economic criteria when making decisions about treatments is an important component of integrated pest management. Economic damage is the point when the loss in value caused by damage to a crop equals the cost of control. The economic injury level is the population at which a pest begins causing economic damage to a crop. The goal of integrated pest management is to use all available methods to keep pest populations below this level, but not to completely eradicate them. The action threshold is the point at which treatment should be applied to prevent the economic injury level from being reached. In forest nurseries the action threshold for many pests is low because the potential for economic damage is high. Treatment is worthwhile because large numbers of seedlings are often involved, or because seed values are high. In some cases, the action threshold may be set at zero, in other words, treatments are applied before any damage is observed. Generally, an action threshold of zero is used when dealing with pests whose populations increase rapidly, so that by the time damage is observable the economic injury level has been reached. Such treatments include fumigation, seed treatments, pre-emergent herbicides, and protectant systemic insecticides and fungicides.

Evaluating Damage After Outplanting

Damage to seedlings caused by pests or abiotic factors, such as freeze injury or anaerobic conditions in the nursery may carry over to affect outplanting success. If seedlings are not available for planting, investments in site preparation may be lost. The site may have to be prepared again when seedlings become available in subsequent years. Insects and diseases originating in the nursery are frequently transported to outplanting sites on infested or infected seedlings. Many of these insects and pathogens do not survive well in the natural environment. However, some are capable of surviving and may continue to damage planted seedlings. Insects and pathogens may also be spread from one nursery to another on seedlings that are shipped between facilities.

Not all outplanting problems are the result of problems that may have occurred in the nursery. Determining whether damage to seedlings observed after outplanting originated in the nursery requires careful examination of signs and symptoms, and often involves retracing the history of the stock from lifting through storage and planting to find clues. Successful diagnosis may be difficult if adequate records have not been kept. Information that is helpful in evaluating outplanting problems includes signs and symptoms observed on damaged seedlings; records from lifting, packing, storage, and transportation (date, duration, location, temperature, and type of container); outplanting site conditions (site index, soil type, site preparation methods including any chemical applications, and competition); planting records (seedling quality, planting methods, planting quality, and weather); and records of subsequent events such as unusual weather, fertilization, and herbicide or other release treatments.

The most serious consequence of outplanting insect-infested or diseased seedlings is the potential for introduction of invasive pests and their subsequent spread to native forests. White pine blister rust and Port-Orford cedar root disease are two examples of diseases that were introduced on nursery stock early in the twentieth century and spread into native forests, causing widespread mortality. More recently, the introduction and spread of *Phytophthora ramorum*, the causal agent of sudden oak death and Ramorum blight, has highlighted the potential for spread of diseases between nurseries, and from nurseries to native forests. It emphasizes the importance of careful evaluation and identification of diseases and insects in nursery stock. Most States have regulations for inspection and certification of nursery stock to protect against interstate movement of *Phytophthora ramorum* and other invasive pests.

Cooperation between nursery managers and landowners or managers in evaluation of damage after outplanting can improve both nursery production and outplanting success. The success of these evaluations depends upon accurate records being kept from sowing to outplanting by the nursery manager and those responsible for outplanting and seedling establishment. Participation in such evaluations can enhance a nursery manager's reputation as a professional with an interest in providing the highest quality seedlings to ensure reforestation and restoration success.

Selected References

Cleary, B.D.; Greaves, R.D.; Hermann, R.K. 1978. Regenerating Oregon's forests: a guide for the regeneration forester. Corvallis, OR: Oregon State University Extension Service. 287 p.

Duryea, M.L.; Landis, T.D., eds. 1984. Forest nursery manual: production of bareroot seedlings. Boston, MA; The Hague, The Netherlands; Lancaster, PA: Martinus Nijhoff/ Dr. W. Junk Publishers. 386 p.

Landis, T.D.; Tinus, R.W.; McDonald, S.E.; Barnett, J.P. 1990. The biological component: nursery pests and mycorrhizae, vol. 5. The container tree nursery manual. Agriculture Handbook 674. Washington, DC: USDA Forest Service: 171 p.

Smith, Jr., R.S.; Cordell, C.E. 1989. Evaluation of nursery losses due to pests. In: Cordell, C.E.; Anderson, R.L.; Hoffard, W.H.; Landis, T.D.; Smith, Jr., R.S.; Toko, H.V., tech. coords. Forest nursery pests. Agriculture Handbook 680. Washington, DC: USDA Forest Service: 14–15.

South, D.B.; Enebak, S.A. 2006. Integrated pest management practices in southern pine nurseries. New Forests. 31: 253–271.

Soil-Pest Relationships

Jerry E. Weiland

Soil is essential for the production of healthy nursery stock. It provides physical support for roots and supplies mineral nutrients and water necessary for growth. The soil is also the environment in which plant roots interact with soilborne insects and pathogens. Therefore, an understanding of how soil properties affect both plant and pest health is critical for making effective pest management decisions.

Soil is a living, dynamic body composed of mineral solids, air, water, and organic matter (including living organisms). Although soil characteristics vary greatly throughout the United States, certain basic soil properties are important in mediating soil-pest relationships. Some properties, such as soil texture, are relatively fixed. Other properties, however, can be modified to favor plant health over that of injurious insects and pathogens. Knowing your site along with its associated soil properties can go a long way in determining which disease and pest management practices will be most appropriate. The most relevant soil properties are discussed below.

Texture

Soil texture (percentages of sand, silt, and clay) is one of the most important properties in soil-pest relationships because it directly affects a number of other soil characteristics crucial for plant growth including nutrient availability, gas exchange, and soil moisture levels. In general, sandy loams or loamy sands are recommended for nursery production because they are more resistant to compaction and provide enough pore space for adequate drainage and air infiltration. Coarse-textured soils are also more workable across a wider range of soil moisture levels and resist large clod formation, thus making soil management easier. For example, fumigants infiltrate farther in sandy soils with large pore spaces and few soil aggregates (clods).

Fine-textured soils are more prone to compaction and susceptible to poor drainage. Both conditions inhibit root growth and result in plant stress that can increase susceptibility to root pathogens. Damping-off and root rot are frequently more severe in clay soils because of these conditions. Fine-textured soils may slow seedling emergence, thereby increasing the amount of time that seedlings are in contact with pathogenic fungi. Garden symphylans (*Scutigerella immaculata*) also tend to be more problematic in heavier soils.

Coarse-textured soils, on the other hand, are more favorable for nematodes, which require larger pore spaces to move easily. For example, the root lesion nematode (*Pratylenchus penetrans*) and *Xiphinema bakeri*, the cause of corky root disease, are both more severe in sandy soils. Root feeding insects, such as the black vine weevil (*Otiorhynchus sulcatus*), strawberry root weevil (*Otiorhynchus ovatus*), and pales weevil (*Hylobius pales*) similarly prefer coarse-textured soils.

Moisture

Soil moisture is another important property that mediates soil-pest relationships. Water is essential for life and its presence directly affects both plant and pest health. Because the amount of water in a soil is inversely related to the amount of soil aeration, maintaining the proper balance of water and air is key to plant health. If too much water is present for an extended period of time, roots can suffocate. However, too little water can lead to drought stress. In either case, roots can sustain damage and become more susceptible to insect and pathogen attack.

Although local climatic conditions and water tables strongly influence soil moisture, appropriate drainage tile placement and the installation of an irrigation system can help mitigate the effects of standing water or a dry nursery site. Soil moisture conditions may also need to be managed for effective pest control. Fumigant efficacy is dependent on soil moisture during and after fumigation, and excess water can lead to pesticide leaching or uneven pesticide distribution in nursery fields.

Two classic examples of disease favored by excess water are damping-off and root rot caused by *Pythium* and *Phytophthora* species. Both pathogen genera produce motile spores that swim to host plant roots when soil moisture is abundant. Certain insects such as the European crane fly (*Tipula paludosa*) and fungus gnats (*Bradysia* and *Lycoriella* spp.) also prefer moist conditions.

Other disease and pest problems are favored by low soil moisture. Damping-off caused by *Rhizoctonia solani* and *Fusarium* species can be more severe under dry conditions. Drought stress may increase plant susceptibility to root rot caused by *Cylindrocladium* species. Similarly, charcoal root rot of pine, caused by *Macrophomina phaseolina*, is much more severe in hot, dry soils. Drought also predisposes hosts to stem diseases such as Phomopsis canker on Douglas-fir, Diplodia shoot blight and canker on pine, and Cytospora canker on hardwoods.

Temperature

Soil temperature affects the rate of seed germination, plant growth, and the development and survival of soil pests. The predominant factors that influence soil temperature are local climatic conditions, soil characteristics, and the amount of vegetative cover shielding the soil from incoming solar radiation. Several nursery practices, however, may be used to moderate soil temperature extremes. In summer, irrigation and mulching can be used to cool nursery soils. Mulching also works to keep the soil warmer during cold weather by reducing heat loss. In areas that receive adequate sun, soil solarization with clear plastic has proven effective in reducing insect and pathogen populations.

Warm soil temperatures are favorable for the development of a number of diseases including charcoal root rot and Fusarium root rot. Warm, moist conditions also increase infection by *Cyclindrocladium* species, which cause root rot in a number of ornamentals. In some situations, excessive soil temperatures predispose seedlings to infection. In the case of Fusarium hypocotyl/root rot on Douglas-fir seedlings, symptoms often appear with the first onset of hot summer weather when roots are unable to supply adequate amounts of water to actively growing seedlings. Warm soil temperatures may also be crucial for biological control efficacy. *Heterorhabditis marelatus*, an insect-feeding nematode, requires consistent soil temperatures above 10 °C (50 °F) to be effective against black vine root weevil.

Cool soil temperatures slow seed germination, seedling growth, and woody tissue development. These conditions prolong the amount of time that seedling tissues remain succulent, and therefore susceptible to infection by damping-off pathogens. Cooler soil temperatures also slow pest development and may reduce the number of generations that occur each year. Garden symphylans take 5 months to complete their life cycle at 10 °C (50 °F), but require less than 2 months at 25 °C (77 °F). In addition, cold soil temperatures restrict the geographic range of certain insects and pathogens. The root pathogens *Phytophthora cinnamomi* and *Sclerotium rolfsii* are both limited by low soil temperatures and do not survive where soils regularly freeze.

Soil Reaction (pH)

Soil pH is a measure of the acidity or alkalinity of the soil. In general, slightly acid soils (pH 5 to 6) are recommended for optimal seedling growth of both hardwoods and conifers. Micronutrients such as iron and manganese become more available in slightly acid soils, but others such as aluminum may become toxic as pH levels decrease. Soil acidity is generally modified through the addition of soil amendments. Liming serves to increase pH and make soils more alkaline. Soil sulfur is commonly used to quickly lower pH, thus promoting soil acidity.

The severity of several diseases may be significantly reduced by soil acidification. Black root rot, caused by *Thielaviopsis basicola*, has been suppressed in soils with a pH less than 5.2. Losses from damping-off caused by *Fusarium*, *Phytophthora*, *Pythium*, and *Rhizoctonia* species have also been reduced by keeping soil pH low. Some acid-tolerant pathogens such as *Cylindrocladium*, however, may actually increase disease severity if the soil is acidified. Aluminum sulfate, a common soil amendment to lower soil pH, has shown some efficacy in reducing disease severity because several soilborne pathogens are sensitive to aluminum and increased soil acidity. However, caution must be used because high aluminum levels can be toxic to plants.

Symptoms of iron, manganese, and zinc deficiency are common on certain plant species grown in soils with a high pH. Pin oak and river birch seedlings, for example, are often chlorotic in alkaline soils. Treatment options are determined after a nutrient analysis of the affected plants and include soil acidification or fertilization with the appropriate nutrient.

Relatively little information exists about the effect of soil pH on nursery insect pests. However, the pH range of most nursery soils is unlikely to limit insect pests known to occur in nurseries.

Nutrients

Optimal plant nutrition is an important component of plant health because it helps maintain plant defenses. As with water, too little or too much of any nutrient can negatively affect plant health and increase susceptibility to insect and pathogen attack. The best strategy is to maintain a balanced fertilization program without promoting excessive growth.

The three main constituents of most fertilizers—nitrogen, phosphorous, and potassium—have been shown to affect disease incidence and severity. Nitrogen is the most studied nutrient and is essential in many physiological processes for growth and disease resistance. Excess nitrogen can lead to succulent growth and delayed tissue maturity, thereby making plants more susceptible to fungal infection. Damping-off, for instance, is frequently more severe after heavy nitrogen fertilization. In contrast, phosphorous decreases the severity of damping-off in conifers by *Fusarium oxysporum*. Phosphorous accelerates tissue maturity, thus increasing resistance to several plant pathogens, particularly those that attack young tissues. Phosphorous also helps with new root generation, which may assist seedlings in outgrowing root damage. Although potassium assists root growth, increased potassium levels in Monterey pine have been associated with greater susceptibility to Phytophthora root rot. Where fusiform rust (*Cronartium fusiforme*) is prevalent, high concentrations of all three nutrients are associated with increased susceptibility and damage.

With the exception of calcium, little work has been done on the effect of other nutrients in mediating soil-pest relationships. Calcium amendments such as gypsum, calcium chloride, and calcium carbonate have reduced infection by *Phytophthora* species in seedlings of fruit trees and ornamentals. It is thought to function by strengthening plant cell walls, thus increasing resistance to infection, and by directly interfering with pathogen biology.

Relatively little research has been conducted on the effect of plant nutrition on root insects of nursery crops. About the only generalization that can be made is that most root feeding insects tend to have long generation times (root weevils 1 to 2 years, cicada up to 17 years, white grubs 2 to 4 years) because the availability of nitrogen in roots is low.

Organic Matter

A significant proportion of research in the last few decades has focused on the role of organic matter (OM) in the management of soilborne diseases. Organic matter benefits the soil by increasing water holding capacity, soil aeration, and nutrient availability. However, the addition of too much carbon by incorporating amendments such as sawdust or bark mulch directly into the soil can result in temporary nitrogen deficiencies. Three strategies of OM management (bare fallow, specific OM amendments, and suppressive soils) have emerged in the past few decades and show promise for pathogen management. The first, bare fallow, is a relatively simple strategy for the reduction of certain soilborne pathogen populations. The latter two strategies, however, are more complicated and will likely require additional research to realize their full potential.

Periodic bare fallow treatments (every third year) have proven effective in reducing pathogen populations in bareroot conifer nurseries. This strategy works by removing the host and by decreasing the amount of OM in the soil, thereby eliminating or decreasing the food base for soilborne pathogens. Soils without vegetative cover also experience greater temperature and moisture extremes that may contribute to pathogen mortality. Significant reductions in soilborne pathogens such as *Pythium*, *Fusarium*, and *Cylindrocarpon* species have been observed and in several cases bare fallow has been as effective as fumigation in reducing pathogen populations. However, bare fallow is less likely to work as well for certain soilborne pathogens, such as *Cylindrocladium* species or *Macrophomina phaseolina*, which produce resistant structures that survive for years in the soil.

Incorporation of specific OM amendments, such as compost or green cover crops, has shown promise in reducing soilborne pathogen populations in nursery soils. Plants in the cabbage (Brassicaceae), onion (Alliaceae), and grass (Poaceae) families have received particular attention because of their ability to produce compounds that are directly toxic to soilborne pathogens. Perhaps the best known example is the use of *Brassica* species as a soil amendment to produce isothiocyanates and other fungitoxic compounds. Although some trials have been promising, overall the results have been inconsistent and much more research needs to be conducted before this method gains acceptance. To further complicate matters, populations of *Fusarium* species have often increased dramatically after OM amendment. Nevertheless, not all isolates of *Fusarium* are pathogenic and it is unknown what effect these population surges have on disease severity.

Suppressive soils are those in which disease incidence or severity is minimal despite the presence of the pathogen and susceptible plant host. Suppressiveness is often a function of the soil microbial community, although the soil's physical and chemical attributes may play a role. The functionality of suppressive soils is maintained through a variety of means, including crop rotation and the addition of OM amendments, in an effort to increase resident soil microbial diversity or to enhance the presence of microbes that are antagonistic to soilborne pathogens. Some microbes identified from suppressive soils compete with soilborne pathogens for resources or directly affect pathogen growth. Others, such as nonpathogenic *Fusarium oxysporum* isolates, have induced systemic resistance in crop plants. Research to date has focused primarily on agronomic and fruit crops, but the knowledge gained from these systems will assist interested nurseries in the development of healthy soils with greater microbial diversity.

Selected References

Anderson, R.L.; Sutherland, J.R. 1989. Soil-pest relationships. In: Cordell, C.E.; Anderson, R.L.; Hoffard, W.H.; Landis, T.D.; Smith, Jr., R.S.; Toko, H.V., tech. coords. Forest nursery pests. Agriculture Handbook 680. Washington, DC: USDA Forest Service: 16–17.

Datnoff, L.E.; Elmer, W.H.; Huber, D.M., eds. 2009. Mineral nutrition and plant disease. St. Paul, MN: APS Press. 278 p.

Duryea, M.L.; Landis, T.D., eds. 1984. Forest nursery manual: production of bareroot seedlings. Boston, MA; The Hague, The Netherlands; Lancaster, PA: Martinus Nijhoff/Dr. W. Junk Publishers. 386 p.

Gugino, B.K.; Idowu, O.J.; Schindelbeck, R.R.; van Es, H.M.; Wolfe, D.W.; Thies, J.E.; Abawi, G.S. 2007. Cornell soil health assessment training manual, edition 1.2.1. Geneva, NY: Cornell University. 52 p.

Hansen, E.M.; Lewis, K.J. 1997. Compendium of conifer diseases. St. Paul, MN: American Phytopathological Society. 101 p.

Hildebrand, D.M.; Stone, J.K.; James, R.L.; Frankel, S.J. 2004. Alternatives to preplant soil fumigation for Western forest nurseries. PNW-GTR-608. Portland, OR: USDA Forest Service, Pacific Northwest Research Station. 27 p.

Höper, H.; Alabouvette, C. 1996. Importance of physical and chemical soil properties in the suppressiveness of soils to plant diseases. European Journal of Soil Biology. 32(1): 41–58.

Johnson, W.T.; Lyon, H.H. 1991. Insects that feed on trees and shrubs. Ithaca, NY: Cornell University Press. 560 p.

Jones, R.K.; Benson, D.M. 2003. Diseases of woody ornamentals and trees in nurseries. St. Paul, MN: APS Press. 482 p.

Parker, C.A.; Rovira, A.D.; Moore, K.J.; Wong, P.T.W.; Kollmorgen, J.F., eds. 1985. Ecology and management of soilborne plant pathogens. St. Paul, MN: APS Press. 358 p.

Sinclair, W.A.; Lyon, H.H. 2005. Diseases of trees and shrubs, 2nd ed. Ithaca, NY: Cornell University Press. 660 p.

Weller, D.M.; Raaijmakers, J.M.; McSpadden Gardener, B.B.; Thomashow, L.S. 2002. Microbial populations responsible for specific soil suppressiveness to plant pathogens. Annual Review Phytopathol. 40: 309–348.

Mycorrhizae in Forest Tree Nurseries

Michelle M. Cram and R. Kasten Dumroese

Mycorrhizae are symbiotic fungus-root associations. The colonization of roots by mycorrhizal fungi can benefit the host by improving nutrient and water uptake. In exchange, the host plant provides the mycorrhizal fungi carbohydrates (carbon) from photosynthesis. A substantial portion of this carbon is ultimately transferred to the rhizosphere and is estimated to account for up to 15 percent of the organic matter in forest soils. Under most environmental conditions, trees and other plants are naturally colonized by mycorrhizal fungi. These mycorrhizal associations are highly complex and dynamic, a result of the great diversity of mycorrhizal fungi, hosts, and terrestrial systems that interact and evolve with changes in hosts and environmental conditions.

Two types of mycorrhizae are found on trees: ectomycorrhizae and arbuscular mycorrhizae (syn. endomycorrhizae). Ectomycorrhizal fungi enter the root between cortex cells and often form a thick mantle outside of the short feeder roots that is visible to the naked eye (fig. I.6). Forest tree species with ectomycorrhizae include pine, firs, spruce, hemlock, oak, hickory, alder, and beech.

Arbuscular mycorrhizal fungi enter the root cells and cannot be seen without the aid of a microscope (fig. I.7). Arbuscular mycorrhizae are especially effective at transferring carbon to soil in the form of glomalin, a sticky glue-like substance that is estimated to provide 30 to 40 percent of the carbon found in soils. Forest tree species with arbuscular mycorrhizae include cedars, cypress, junipers, redwoods, maple, ash, dogwoods, sycamore, yellow-poplar, and sweetgum. Agricultural crops used by forest nurseries as cover crops also form arbuscular mycorrhizae.

The intensive commercial seedling production for reforestation typically suppresses or delays colonization of seedlings by mycorrhizal fungi. Fumigation used to control pests can limit and sometimes remove mycorrhizal fungi from the upper 15 to 30 cm (6 to 12 in) of soil for weeks to several months. Some fungicides have also been found to suppress mycorrhizae development, especially systemic fungicides (for example, triadimefon) and fungicides applied as soil drenches (for example, azoxystrobin, iprodione). High fertilization rates have also been known to suppress mycorrhizae development, particularly when the fertilizer is high in phosphorus (greater than 150 ppm). Despite the negative effect these common nursery treatments have on mycorrhizae, the benefits of fertilization and pest control often outweigh the delay of mycorrhizae formation on seedlings.

Figure I.6—Pisolithus tinctorius *forms a yellow-brown (ocher) mantle on pine feeder roots.* Photo by Michelle M. Cram, USDA Forest Service.

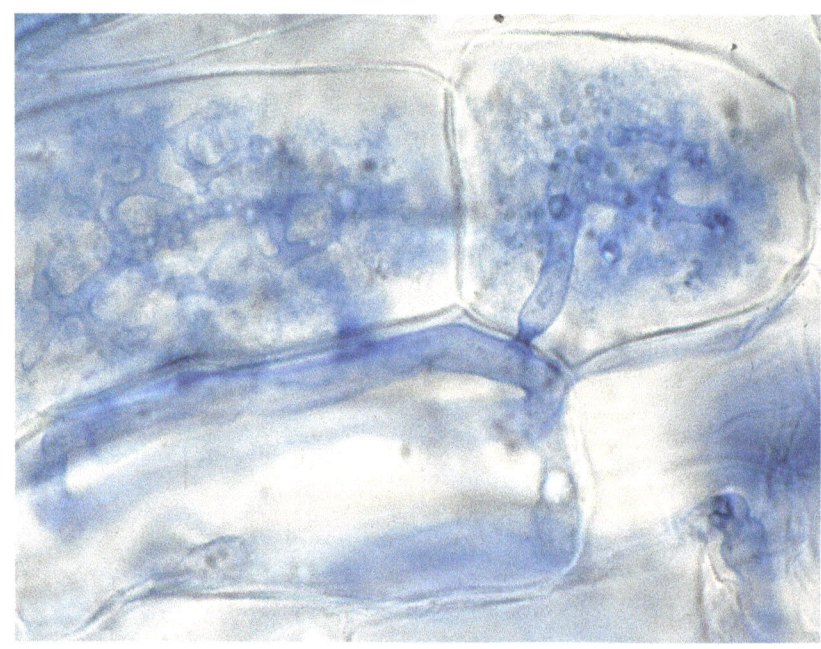

Figure I.7—*Arbuscular mycorrhizal fungi within western redcedar root cells.* Photo by Michael P. Amaranthus, Mycorrhizal Applications, Inc.

Ectomycorrhizal fungi are able to infest fumigated fields or containers via windblown spores produced by mushrooms and puffballs growing within or near the nurseries. These ectomycorrhizal fungi are often well-adapted to intensive nursery systems. *Thelephora terrestris* is common in most North American nurseries (fig. I.8). In the Pacific Northwest, *Laccaria laccata* and *Inocybe lacera* are also common forest nursery colonizers. In contrast, arbuscular mycorrhizal fungi produce soilborne spores that are unlikely to colonize container growing media and can be slower to recolonize fumigated soils. This delay in colonization can limit production of species highly dependent on arbuscular mycorrhizae.

Artificial inoculation with mycorrhizal fungi can be beneficial for a variety of tree seedlings and sites; however, there are many documented cases in which artificial inoculation resulted in no measurable benefit or significant losses in seedling survival and growth. The symbiotic association between plants and mycorrhizal fungi requires a balanced exchange of carbon to expanded nutrient and water uptake to be mutually beneficial. If the environmental conditions that plants are growing in are fully adequate for growth, then mycorrhizal fungi may add little benefit to offset their carbon use. Rapid colonization of seedlings by mycorrhizal fungi naturally present in the nursery or at a typical outplanting site may also render artificial inoculation unnecessary. Results of artificial inoculation will vary greatly depending on the host, species or strain of mycorrhizal fungi, and environmental conditions of the nursery and outplanting sites. The complexity of interactions between mycorrhizal associations and environmental conditions requires careful testing of a particular mycorrhizal inoculant to ensure a positive benefit-to-cost ratio before operational use.

Artificial inoculation with mycorrhizal fungi in the nursery is used to increase seedling performance in situations known by researchers and managers to have consistently positive results. *Pisolithus tinctorius*, an ectomycorrhizal fungus, has been known to increase survival and growth of pine and oak seedlings on strip mine spoils in the Eastern United States (fig. I.9). Artificial inoculation with arbuscular mycorrhizal fungi (for example, *Glomus intraradices*) for species such as incense cedar, redwood, giant sequoia, and western redcedar significantly increase seedling density and growth in the nursery, and survival and early growth after outplanting on some sites. Many other examples exist of nurseries that use artificial inoculation of mycorrhizal fungi in hardwood beds following fumigation, container operations, or as requested by their clients.

Mycorrhizal fungi may be purchased, usually in the form of spore inoculum. These products typically are applied in the nursery as a dry granular to soil or media before sowing or as a drench after germination. Some container media are sold with mycorrhizal fungi or spores incorporated. It may be difficult to find commercial products that contain only one species of mycorrhizal fungi. Many products often combine multiple species of mycorrhizal fungi, other biological organisms (for example, bacteria and *Trichoderma* spp.), and nutrients, making each ingredient's benefit difficult to assess. When testing mycorrhizal products that also contain nutrients, it is imperative to include a nutrient control to determine if seedling response is caused by the fungi

Figure I.8—Thelephora terrestris *fruiting bodies on slash pine.* Photo from USDA Forest Service Archive.

Figure I.9—Pisolithus tinctorius (Pt) *fruiting bodies form under a Virginia pine originally inoculated with Pt and planted on a strip-mine spoil.* Photo by Michelle M. Cram, USDA Forest Service.

or the nutrients. Managers interested in using or testing a particular mycorrhizal species may have to special order a single species product.

Some nurseries produce their own mycorrhizal inoculum. Ectomycorrhizal inoculum is made by grinding up the fruiting bodies (puffballs, truffles, etc.) and adding the spores to soil, dusting seeds, or mixing the inoculum with water and drenching containers or beds after germination (fig. I.10). Nurseries that use pesticides known to suppress mycorrhizae should delay artificial inoculation until after the pesticide applications are finished. The primary limitation to producing ectomycorrhizal inoculum is finding sufficient quantities of fruiting bodies. In contrast, arbuscular mycorrhizal inoculum can be grown in pots with fast growing host plants. Original inoculum is collected in the field from soil under desired tree species and mixed with container media before sowing host seeds, such as alfalfa and grasses. After the host plants have matured and are well infected (assistance from university or State extension services may be required to determining the presence and species), the soil and host roots can be cut up and added to container media (10 percent by volume) or applied to beds before sowing.

Common cover crops used in bareroot forest nurseries for 1- to 2-year rotations can help boost arbuscular mycorrhizae populations and increase organic matter content. This increase in mycorrhizal fungi, however, will most likely be lost if fumigation occurs between cover cropping and seedling production. Many nurseries that fumigate in the fall will use a winter cover crop, such as rye grass and oats, as living mulch. Winter cover crops in other agriculture systems increase arbuscular mycorrhizae levels in the subsequent summer crops. In forest tree nurseries, more information is needed on the interactions of mycorrhizae, cover crop, tree species, and application timing to deploy a rotational cropping system for managing seedlings dependent on arbuscular mycorrhizae.

The potential of mycorrhizae to positively affect seedlings survival and growth will continue to draw efforts at using artificial and cultural techniques to produce a superior mycorrhizal seedling. A better understanding of the mycorrhizal system for each nursery, tree species, and outplanting site is needed to determine the best cultural or artificial inoculation practices. Ultimately, any practices used in forest nurseries must increase seedling performance and have an acceptable benefit-to-cost ratio.

Selected References

Allen, M.F.; Swenson, W.; Querejeta, J.I.; Egerton-Warburton, L.M.; Treseder, K.K. 2003. Ecology of mycorrhizae: a conceptual framework for complex interactions among plants and fungi. Annual Review of Phytopathology. 41: 271–303.

Amaranthus, M.; Steinfeld, D. 2005. Arbuscular mycorrhizal inoculation following biocide treatment improves *Calocedrus decurrens* survival and growth in nursery and outplanting sites. In: Dumroese, R.K.; Riley, L.E.; Landis, T.D., tech. coords. National proceedings: Forest and Conservation Nursery Associations–2004. Gen. Tech. Rep. RMRS-P-35. Fort Collins, CO: USDA Forest Service, Rocky Mountain Research Station: 103–108.

Barnhill, M.A. 1981. Endomycorrhizae in some nursery-produced trees and shrubs on a surface-mined area. Tree Planters' Notes. 32(1): 20–22.

Figure I.10—*Puffball fruiting body of* Pisolithus tinctorius. Photo by Michelle M. Cram, USDA Forest Service.

Corkidi, L.; Allen, E.B.; Merhaut, D.; Allen, M.F.; Downer, J.; Bohn, J.; Evans, M. 2005. Effectiveness of commercial mycorrhizal inoculants on the growth of *Liquidambar styraciflua* in plant nursery conditions. Journal of Environmental Horticulture. 23: 72–76.

Corkidi, L.; Evans, M.; Bohn, J. 2008. An introduction to propagation of arbuscular mycorrhizal fungi. Native Plants Journal. 9: 29–38.

Corrêa, A.; Strasser, R.J.; Martins-Loução, M.A. 2006. Are mycorrhiza always beneficial? Plant and Soil. 279: 65–73.

Cram, M.M.; Mexal, J.G.; Souter, R. 1999. Successful reforestation of South Carolina sandhills is not influenced by seedling inoculation with *Pisolithus tinctorius* in the nursery. Southern Journal of Applied Forestry. 23: 46–52.

Diedhiou, P.M.; Oerke, E.C.; Dehne, H.W. 2004. Effects of the strobilurin fungicides azoxystrobin and kresoxim-methyl on arbuscular mycorrhiza. Journal of Plant Diseases and Protection. 111: 545–556.

Dosskey, M.G.; Linderman, R.G.; Boersma, L. 1990. Carbon-sink stimulation of photosynthesis in Douglas-fir seedlings by some ectomycorrhizas. New Phytologist. 115: 269–274.

Kabir, Z.; Koide, R.T. 2002. Effect of autumn and winter mycorrhizal cover crops on soil properties, nutrient uptake and yield of sweet corn in Pennsylvania, USA. Plant and Soil. 238: 205–215.

Kormanik, P.P.; Bryan, W.C.; Schultz, R.C. 1980. Increasing endomycorrhizal fungus inoculum in forest nursery soil. Southern Journal of Applied Forestry. 4: 151–153.

Kormanik, P.P.; Schultz, R.C.; Bryan, W.C. 1982. The influence of vesicular-arbuscular mycorrhizae on the growth and development of eight hardwood tree species. Forest Science. 28: 531–539.

Landis, T.D.; Amaranthus, M.A. 2009. Inoculate with mycorrhizae, rebuild your soil, and help stop global warming. Forest Nursery Notes. 29(1): 13, 16.

MacFall, J.S.; Slack, S.A. 1991. Effects of *Hebeloma arenosa* on growth and survival of container-grown red pine seedlings (*Pinus resinosa*). Canadian Journal of Forest Research. 21: 1459–1465.

Marx, D.H.; Artman, J.D. 1979. *Pisolithus tinctorius* ectomycorrhizae improve survival and growth of pine seedlings on acid coal spoils in Kentucky and Virginia. Reclamation Review. 2: 23–31.

Marx, D.H.; Cordell, C.E.; Kormanik, P. 1989. Mycorrhizae: benefits and practical application in forest tree nurseries. In: Cordell, C.E.; Anderson, R.A.; Hoffard, W.H.; Landis, T.D.; Smith, Jr., R.S.; Toko, H.V., tech. coords. Forest nursery pests. Agriculture Handbook 680. Washington, DC: USDA Forest Service: 18–21.

Snyder, C.S.; Davey, C.B. 1986. Sweetgum seedling growth and vesicular-arbuscular mycorrhizal development as affected by soil fumigation. Soil Science Society of America Journal. 50: 1047–1051.

South, D. 1977. Artificial inoculation of fumigated nursery beds with endomycorrhizae. Tree Planters' Notes. 28(3): 3–5, 31.

Teste, F.P.; Schmidt, M.G.; Berch, S.M.; Bulmer, C.; Egger, K.N. 2004. Effects of ectomycorrhizal inoculants on survival and growth of interior Douglas-fir seedlings on reforestation sites and partially rehabilitated landings. Canadian Journal of Forest Research. 34: 2074–2088.

Walker, R.F.; West, D.C.; McLaughlin, S.B.; Amundsen, C.C. 1989. Growth, xylem pressure potential, and nutrient absorption of loblolly pine on a reclaimed surface mine as affected by an induced *Pisolithus tinctorius* infection. Forest Science. 35: 569–581.

West, H.M.; Fitter, A.H.; Watkinson, A.R. 1993. The influence of three biocides on the fungal associates of the roots of *Vulpia ciliata* ssp. *ambigua* under natural conditions. Journal of Ecology. 81(2): 345–350.

Pesticide Regulations

John W. Taylor, Jr.

The pesticides available to nursery managers are constantly changing; new products enter the market while others depart because of new formulations, changes in market share, or new regulations. Currently, Federal pesticide regulation is found primarily in three acts—the Federal Insecticide, Fungicide, and Rodenticide Act (FIFRA) of 1972, as amended; the Resource Conservation and Recovery Act (RCRA) of 1976; and the Comprehensive Environmental Response, Compensation, and Liability Act (CERCLA) of 1980.

Federal Insecticide, Fungicide, and Rodenticide Act

This law defines the conditions for developing, registering, and using pesticides and has been amended several times. The 1978 amendment produced significant changes in the requirements for registration, classification, and use of pesticides. The requirements for registration mainly affect pesticide manufacturing companies and will not be addressed here. The 1978 amendment gave States broader authority and responsibility for registering pesticides. States automatically have authority to register pesticides for use within the State for special local needs. Formerly, States had registration authority only with approval of the U.S. Environmental Protection Agency (EPA). Now, however, EPA may disapprove State registration if the use differs from Federal registration, if it creates an imminent health hazard, or if the State has authorized the use on crops where EPA has not established adequate crop tolerances or residue levels. Two other separate, but closely related, areas in FIFRA have also been amended—classification of pesticides and applicator certification. To make maximum use of the State programs for training and certification of applicators of restricted-use pesticides, EPA is authorized to classify pesticide uses by regulation as a separate part of the registration process. Designation of restricted-use pesticides supports State applicator certification programs, in which almost all States train and certify applicators of restricted-use pesticides. Restriction of pesticide use also provides EPA an alternative way to reduce pesticide risks besides outright cancellation of an applicator's registration. The provisions of FIFRA are legally binding. Violations of the provisions in the act carry both civil and criminal penalties. Private applicators who violate the act are subject to civil or criminal penalties of up to $1,000. Commercial applicators are liable for civil penalties up to $5,000 and criminal penalties of up to $25,000 and 1 year in prison. FIFRA explicitly gives States the primary responsibility for enforcing requirements of the act. FIFRA also requires that commercial applicators keep records on their employment of restricted-use pesticides. The content and length of time the application records must be kept vary from State to State, but 2 years is generally the maximum that records must be kept. State enforcement agencies have the authority to make unannounced visits during normal business hours to inspect the records of commercial applicators.

The 1978 amendments also clarified an issue of broad concern—the legality of pesticide uses or practices not addressed in the label direction. The following pesticide-use practices have been specifically excluded from the definition of "use inconsistent with the label" and are, therefore, permissible:

1. Application of the pesticide at less than the labeled dosage, concentration, or frequency.

2. Application of the pesticide to control an unnamed target pest, as long as the crop, animal, or site is included on the label.

3. Pesticide applied using methods not specifically prohibited by the label wording.

4. Application of pesticide(s) with fertilizers, if not prohibited by the label.

This act was amended by the Food Quality Protection Act (FQPA) of 1996, which sought to resolve inconsistencies between the two major pesticide statutes, FIFRA and the Federal Food Drug and Cosmetic Act (FFDCA). Historically, FIFRA was used as a guide for the registration of pesticides in the United States and prescribes labeling and other regulatory requirements. FIFRA is designed to prevent unreasonable adverse effects on human health or the environment, whereas the FFDCA guides the establishment of tolerances (maximum legally permissible levels) for pesticide residues in food. The primary objectives of the FQPA are—

1. Mandate a single, health-based standard for all pesticides in all foods.

2. Provide special protection for infants and children.

3. Expedite approval of "safer" pesticides.

4. Create incentives for the development and maintenance of effective crop protection tools for American farmers.

5. Require periodic reevaluation of pesticide registrations and tolerances.

Although the FQPA was originally focused on food use pesticides, the risks associated with each pesticide were evaluated across all registered uses of each pesticide, and across all pesticides with a "similar" mode of action, and a "risk cup" which represented the total exposure allowed from all these sources was developed. If the "risk cup" "overflowed," then the registrant, in order to maintain registration of the material, had to reduce the risk until it met the EPA's acceptable level. This is the point at which minor use—nonfood pesticides such as those used in forest tree nurseries—became involved as registrations were removed to reduce composite risk. Implementation of the requirements has significantly affected, and will continue to affect, the range of pesticide choices available to manage pests in forest tree nurseries.

Resource Conservation and Recovery Act

The second Federal regulation governing pesticide use is the Resource Conservation and Recovery Act of 1976. RCRA is designed to extensively monitor hazardous wastes. The act identifies solid wastes that are hazardous and sets forth requirements that govern their handling, storage, and disposal. This law applies to all hazardous wastes unless they are specifically exempted. The EPA administers the act with assistance from the Materials Transportation Bureau of the U.S. Department of Transportation. The major provisions of the act for controlling hazardous wastes include—

1. A definition of "hazardous waste."
2. A manifest system to monitor hazardous waste from its generation to its disposal. This system is frequently called "cradle to grave tracking."
3. A permit requirement for facilities that treat, store, or dispose of hazardous waste.
4. A requirement that every State must have a hazardous waste program.

Any installation that stores, disposes of, transports, or offers to transport hazardous waste must obtain an EPA identification number from the Regional EPA Administrator. Once the "waste" designation has been applied to any pesticide on EPA's hazardous waste list or residues of these pesticides resulting from a spill cleanup, they can be stored for only 90 days before disposal. Otherwise, a storage permit must be obtained from the Regional EPA Administrator. People transporting hazardous wastes for offsite disposal must prepare a shipping manifest that contains certain specific information as described in Section 3010 of RCRA.

Specific pesticide waste regulations were developed by EPA in the Code of Federal Regulations (CFR 260-266 and 122-124) and became effective in October 1980. Pesticides listed in subpart D of Part 261 of the act are not classified as hazardous wastes until the decision is made to dispose of them or they are stored pending disposal. Pesticides that have been determined to be excess are still considered pesticides. Pesticide containers that have been triple rinsed are not considered hazardous wastes under these regulations and can be disposed of either at approved landfills or by burial.

Comprehensive Environmental Response, Compensation, and Liability Act

The Comprehensive Environmental Response, Compensation, and Liability Act established what is known as the "Superfund" for cleaning toxic wastes. This law was enacted in 1980 and is codified in Title 42 of the United States Code, beginning in section 96D.

The Superfund, initially $1.6 billion, was set up to pay for cleaning up spills of hazardous materials and the disposal areas themselves. The act specifies who is liable for reimbursing the fund. It also requires that spills be reported.

There are some basic differences between RCRA and CERCLA. RCRA is not self-implementing. Rather than telling companies what they must do, Congress directed EPA to formulate regulations that would control company activities. CERCLA's approach is very different—it establishes liabilities and obligations and does not require promulgation of regulations to be effective. A second important difference between the two bills relates to the time periods they are designed to control. Whereas CERCLA is designed primarily to determine liability arising out of historic waste management practices, RCRA is designed to affect current hazardous waste management activities. Another difference is that although the principal burden of implementing CERCLA rests with the Federal Government, the requirements of RCRA are primarily a State responsibility, subject to EPA approval.

These laws affect nurseries when pesticides classified as hazardous are disposed (RCRA), or when problems arise (CERCLA). Nursery managers who have questions regarding this subject can contact the hazardous waste management officials in their State. Pesticides may be used safely and effectively to control nursery pests, but careful attention must be paid to both the Federal and State laws governing their use. Questions regarding the legal use of pesticides should be directed to the proper pesticide control official.

State Pesticide Regulations

In addition to the three Federal laws, States also have regulations governing the purchase, use, and disposal of pesticides. Before any pesticide use project is initiated in any State, a responsible official must determine that all applicable State pesticide regulations are being followed.

Selected References

Taylor, Jr., J.W. 1989. Pesticide regulations. In: Cordell, C.E.; Anderson, R.L.; Hoffard, W.H.; Landis, T.D.; Smith, Jr., R.S.; Toko, H.V., tech. coords. Forest nursery pests. Agriculture Handbook 680. Washington, DC: USDA Forest Service: 22–23.

Conifer Diseases

Conifer Diseases

1. Brown Spot Needle Blight
Scott A. Enebak and Tom Starkey

Revised from chapter by Albert G. Kais, 1989.

Hosts

Brown spot needle blight, caused by the fungus *Mycosphaerella dearnessii* (syn. *Scirrhia acicola, Eruptio acicola*), is most commonly found on longleaf pine. The fungus also infects seedlings of slash, loblolly, shortleaf, spruce, pitch, pond, Sonderegger (longleaf x loblolly pine), Virginia, Scots, and eastern white pine.

Distribution

Brown spot needle blight occurs in nurseries throughout the Southern United States. It has also been reported on susceptible pines in plantations, landscapes, and Christmas tree plantations in the Central Plains and Great Lakes regions, Oregon, and Vermont.

Damage

Infected seedlings are seldom killed in the nursery, but repeated and severe infections result in defoliation that reduces seedling vigor, which, in turn, may result in poor seedling survival and growth following outplanting. Longleaf pine seedlings are most susceptible while in the grass stage as conditions near the ground favor the pathogen and seedling infection. The buildup of fungal inoculum increases the number of infections on needles that eventually coalesce and result in defoliation (fig. 1.1). Generally, after the seedling is out of the grass stage (1 m or 3.2 ft), the disease is not an issue. On some pine species used for Christmas trees, however, the disease can cause significant economic losses.

Figure 1.1—*Longleaf pine seedlings infected with brown spot needle blight in the nursery.* Photo by Edward L. Barnard, Florida Division of Forestry.

Diagnosis

Infections and the appearance of lesions occur from May to October. Infection begins with small, grayish-green spots, which become a straw-yellow color and then light brown with chestnut-brown margins (fig. 1.2). Spots coalesce, and the needle tissue dies beyond and between infection sites. Needles with multiple lesions appear mottled and have three distinct zones: a green basal portion, a mottled middle portion, and a dead apical portion (fig. 1.3). Severe infection results in premature defoliation or needle cast and reduced photosynthetic surface and seedling vigor, which prolongs the grass stage and could eventually affect seedling survival.

Two types of fruiting bodies are produced. (1) Conidia are produced in acervuli, which appear on lesions as small black dots visible to the naked eye (fig. 1.4). Conidia are exuded in sticky masses up to 1 mm long that split the epidermis of the leaf. These spores are cylindrical, curved,

Figure 1.2—*Typical lesions caused by* Mycosphaerella dearnessii *on conifer needles.* Photo from USDA Forest Service Archive.

Conifer Diseases

1. Brown Spot Needle Blight

Figure 1.3—*Longleaf pine seedling with advanced symptoms of brown spot needle blight.* Photo by George Blakeslee, University of Florida.

Figure 1.4—*Fruiting bodies (acervuli) of* Mycosphaerella dearnessii *on needle.* Photo by Edward L. Barnard, Florida Division of Forestry.

1 to 4 septate, olive green to brown, and 19 to 35 by 3.5 to 4 microns (fig. 1.5). (2) Ascospores are produced in pseudothecia embedded in dead leaf tissue. They are hyaline, oblong-cuneate, unequally two-celled, and 15 to 19 by 3.5 to 4.5 microns, with two prominent oil drops in each cell (fig. 1.6). Brown spot needle blight can be mistaken for Dothistroma needle blight as the two diseases have similar symptoms and fruiting bodies.

Biology

Seedling infection occurs when spores enter the needles through the stomata. Ascospores are released during periods of high moisture (rain, dew, and fog), can be disseminated great distances by the wind, and are the principal means by which the fungus infects seedling nursery beds. Conidia are produced in acervuli on needle lesions that result from ascospore infection. In contrast to the long-distance movement of ascospores, conidia are disseminated short distances by rain splash

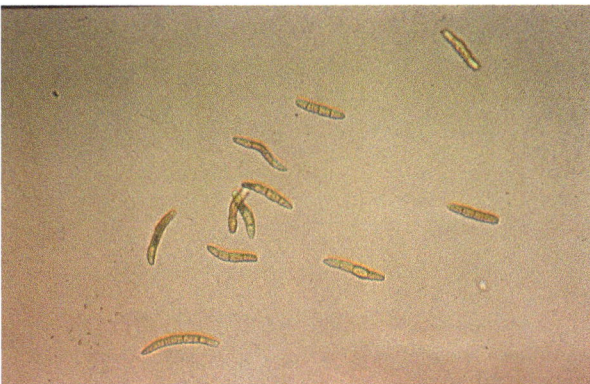

Figure 1.5—*Conidia of* Mycosphaerella dearnessii. Photo by Albert Kais, USDA Forest Service, at http://www.bugwood.org.

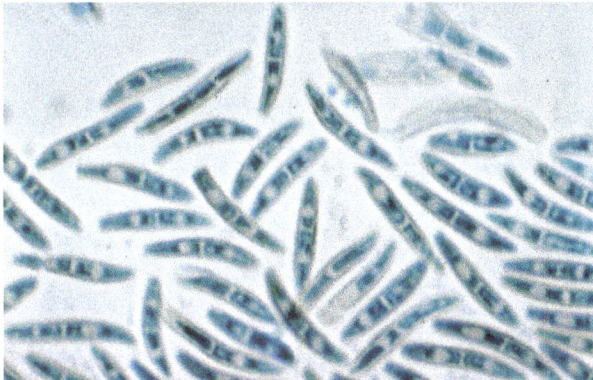

Figure 1.6—*Ascospores of* Mycosphaerella dearnessii. Photo by H.C. Evans, CAB International, at http://www.bugwood.org.

Conifer Diseases

1. Brown Spot Needle Blight

and cause local buildup of the disease in seedling beds. Ascospores are produced on seedlings 2 to 3 months after the seedlings are initially infected. Both spore forms overwinter in dead and infected needle tissue.

Control

Prevention

Inherent susceptibility differences to *M. dearnessii* exist in longleaf pine. The use of longleaf seed sources known to be resistant to infection will decrease the chance of infection and disease in the nursery and after outplanting.

Cultural

Sowing seed in bareroot nurseries at densities of 160 seedlings per m^2 (15 seedlings per ft^2) or less will increase ventilation among seedlings. Mulch is used to reduce mortality from sand splash. Top-clipping longleaf pine needles during the growing season will decrease moisture, decrease conditions necessary for infection, and increase fungicide efficacy. Removing clipped needles from nursery beds decreases the level of inoculum if the fungus is present. This decrease may be important if longleaf pine will be sown in the same area during the next cropping cycle.

Chemical

The most effective control measure is the application of fungicides. Careful monitoring of the seedling crop should be conducted during the growing season. When symptoms first appear, fungicides labeled for brown spot needle blight control should be applied.

Selected References

Evans, H.C. 1984. The genus *Mycosphaerella* and its anamorphs *Cercospetoria*, *Dothistroma* and *Lecanosticta* on pines. Commonwealth Mycological Institute. Mycological Paper 153. 102 p.

Griggs, M.M.; Schmidt, R.A. 1986. Disease progress of *Scirrhia acicola* in single and mixed family plantings of resistant and susceptible longleaf pine. In: Peterson, G.W. Recent research on conifer needle diseases: Proceedings 1984. GTR-WO-50. Gulfport, MS: USDA Forest Service. 106 p.

Jewell, Sr., F.F. 1983. Histopathology of the brown spot fungus on longleaf pine seedlings. Phytopathology. 73: 854–858.

Kais, A.G. 1975. Environmental factors affecting brown spot infection on longleaf pine. Phytopathology. 65: 1389–1392.

Kais, A.G. 1989. Brown spot needle blight. In: Cordell, C.E.; Anderson, R.L.; Hoffard, W.H.; Landis, T.D.; Smith, Jr., R.S.; Toko, H.V., tech. coords. Forest nursery pests. Agriculture Handbook 680. Washington, DC: USDA Forest Service: 26–28.

Snyder, E.B.; Derr, H.J. 1972. Breeding longleaf pine seedling resistance to brown spot needle blight. Phytopathology. 62: 325–329.

Conifer Diseases

2. Cylindrocarpon Root Disease
Robert L. James

Hosts

Cylindrocarpon species have very wide host ranges, including many agricultural, natural, and nursery plant species. In forest nurseries, the major affected species include five-needle pines, especially western white pine and whitebark pine, Douglas-fir, true fir, and occasionally other pine species and western larch.

Distribution

Cylindrocarpon-associated diseases probably occur at some level in most bareroot and container nurseries in the Western United States. Most damage has been associated with bareroot nurseries in the Pacific Northwest and container nurseries in both inland and coastal areas.

Damage

Damage on container white pine and whitebark pine can be substantial within container nurseries (fig. 2.1). *Cylindrocarpon* may cause severe root decay, with epidermal and cortical tissues in both the primary and secondary roots being affected. Although diseased container seedlings may sometimes appear dwarfed, aboveground symptoms are not often evident. Root decay can be extensive, however, and is noticed after seedlings are lifted from containers. Although seedlings with severe root decay must be culled, those that are slightly affected usually perform satisfactorily after outplanting on forest sites. *Cylindrocarpon* is often rapidly replaced with other, nonpathogenic fungi after outplanting.

Container or bareroot transplants may become seriously diseased by *Cylindrocarpon* species after transplanting in nursery soils. Affected seedlings often have chlorotic or necrotic foliage, bottle brush appearance of recent growth, and blackened or decayed roots. Severely affected seedlings are killed.

Figure 2.1—*Foliar symptoms associated with Cylindrocarpon root disease.* Photo by Robert L. James, USDA Forest Service.

Diagnosis

Typical root disease symptoms may not occur on container stock with extensive root decay induced by *Cylindrocarpon*. Root decay becomes evident after lifting diseased seedlings from containers. On bareroot and transplant stock, typical root disease symptoms are common, including foliar chlorosis and necrosis, reduced growth, black or decayed roots, and mortality. Other nursery pathogens, however, species such as *Fusarium, Pythium,* and *Phytophthora,* may cause similar symptoms. Therefore, isolations from diseased seedling roots onto selective media are required for pathogen identification.

Biology

The most commonly isolated *Cylindrocarpon* species from diseased conifer seedlings is *C. destructans*. Other species, however, including *C. didymum, C. tenue*, and *C. cylindroides*, are sometimes associated with disease. *Cylindrocarpon* species are commonly soilborne and are especially prevalent within the rhizosphere of many plants. At least three spore types are produced by most *Cylindrocarpon* species: multicelled macroconidia (fig. 2.2), one- or two-celled microconidia, and thick-walled resting spores called chlamydospores. Macroconidia and microconidia are produced from phialides located on various branched conidiophore types. Chlamydospores form within plant tissues following nutrient extraction and within macroconidia. When susceptible hosts are present, carbohydrate exudates from roots stimulate chlamydospore germination in soil. *Cylindrocarpon* rapidly colonizes the

Conifer Diseases

2. Cylindrocarpon Root Disease

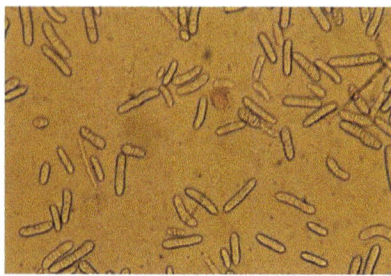

Figure 2.2—*Multicelled macroconidia typical of those produced by* Cylindrocarpon *species.* Photo by Robert L. James, USDA Forest Service.

rhizosphere, invades cortical and vascular tissues, and induces root decay. In some cases, *C. destructans* can induce cortical tissue necrosis without actually invading cortical cells by producing plant toxins. After colonization, roots are decayed or host plants are killed, chlamydospores form and the cycle is repeated. Although some *Cylindrocarpon* species have perfect (sexual) states, which may produce other spores called ascospores, the imperfect (*Cylindrocarpon*) state is the most important for disease epidemiology.

Control

Cultural

Roots from a previous susceptible seedling crop can be an important *Cylindrocarpon* inoculum source in nursery soil. By rotating a nonsusceptible crop into infested soils or by fallowing for at least one growing season (accompanied by periodic fallowed soil tilling), damage from *Cylindrocarpon* can be decreased. In container nurseries, *Cylindrocarpon* species are usually most damaging to over-watered seedlings. Therefore, proper irrigation regulation and the use of well-drained growing media are both important in limiting damage to container stock.

Chemical

Preplant soil fumigation reduces or eliminates potentially pathogenic fungi in nurseries, including *Cylindrocarpon*. Most fumigants that are effective against other soilborne pathogens adequately control *Cylindrocarpon*. *Cylindrocarpon* can be introduced into nursery soils on infected transplant roots; this introduction may be especially serious if transplanting into fumigated soil because few other organisms remain after fumigation to limit pathogen inoculum buildup. Careful transplant culling for root decay and fungicide root dips prior to transplanting may help limit disease severity. Drenching soil with fungicides is usually ineffective in seedbeds or transplant fields after disease symptoms become evident.

Cylindrocarpon can be introduced into container stock on reused containers that have been inadequately sterilized prior to using for a new seedling crop. Hot water or chemical (bleach, copper, and sodium metabisulfite) treatment is necessary for reducing pathogen inoculum on reused containers.

Selected References

Beyer-Ericson, L.; Damm, L.E.; Unestam, T. 1991. An overview of root dieback and its causes in Swedish forest nurseries. European Journal of Forest Pathology. 21: 439–443.

Booth, C. 1966. The genus *Cylindrocarpon*. Commonwealth Mycological Institute. Kew, Surrey, England. Mycological Papers No. 104. 56 p.

Dumroese, R.K.; James, R.L.; Wenny, D.L. 1993. Sodium metabisulfite reduces fungal inoculum in containers used for conifer nursery crops. Tree Planters' Notes. 44(4): 161–165.

Dumroese, R.K.; James, R.L.; Wenny, D.L. 2000. An assessment of *Cylindrocarpon* on container western white pine seedlings after outplanting. Western Journal of Applied Forestry 15(1): 5–7.

Dumroese, R.K.; James, R.L.; Wenny, D.L. 2002. Hot water and copper coatings in reused containers decrease inoculum of *Fusarium* and *Cylindrocarpon* and increase Douglas Fir seedling growth. Horticultural Science 37: 943–947.

Hildebrand, D.M.; Stone, J.K.; James, R.L.; Frankel, S.J. 2004. Alternatives to preplant soil fumigation for western forest nurseries. Gen. Tech. Rep. PNW-GTR-608. USDA Forest Service, Pacific Northwest Research Station. 27 p.

James, R.L. 1989. Effects of fumigation on soil pathogens and beneficial microorganisms. In: Landis, T.D., tech. coord. Proceedings: Intermountain Forest Nursery Association meeting. GTR-RM-184. USDA Forest Service, Rocky Mountain Research Station: 29–34.

James, R.L. 1991. *Cylindrocarpon* root disease of container-grown whitebark pine seedlings—USDA Forest Service Nursery, Coeur d'Alene, Idaho. Report 91-8. USDA Forest Service, Northern Region, Forest Pest Management. 10 p.

James, R.L.; Dumroese, R.K.; Wenny, D.L. 1994. Observations on the association of *Cylindrocarpon* spp. with diseases of container-grown conifer seedlings in the inland Pacific Northwest of the United States. In: Perrin, R.; Sutherland, J.R., eds. Diseases and insects in forest nurseries. Dijon, France, October 3–10, 1993. Institut National De La Recherche Agronominque. Les Colloques No. 68: 237–246.

James, R.L.; Gilligan, C.J. 1990. Root decay of container-grown western white pine seedlings—Plum Creek Nursery, Pablo, MT. Report 90-10. USDA Forest Service, Northern Region, Forest Pest Management. 18 p.

Conifer Diseases

3. Diplodia Shoot Blight, Canker, and Collar Rot
Glen R. Stanosz

Hosts

Diplodia pinea and *D. scrobiculata* (previously known as the single species *Sphaeropsis sapinea*) cause serious diseases of many conifers. Although many conifers, including species of spruce, fir, and larch, can be hosts, severe damage is most common on hard (two- and three-needled) pines, including Austrian, jack, Monterey, mugo, ponderosa, red, and Scots pines.

Distribution

These pathogens are widely distributed in the continental United States, although *D. pinea* appears to be more common and more frequently associated with severe damage.

Damage

Seedlings of all ages may be rendered unmerchantable due to shoot blight, canker, and collar rot leading to deformity or death. In addition, these pathogens may persist on or in asymptomatic seedlings, be transported with seedlings to field sites, and proliferate to cause seedling mortality after planting. Seed rot and damping-off of young seedlings have also been attributed to *Diplodia* pathogens.

Diagnosis

Infection during the first growing season can result in rapid seedling mortality with retention of dead needles (fig. 3.1). On older seedlings, diseased needles often turn yellow, then from red to brown or gray, with stem curling or crooking resulting from shoot death before full needle elongation (fig. 3.2). Cankers on seedling stems begin as discrete, purplish, resinous lesions that result from direct infection or pathogen growth into stems from diseased needles. Collar rot symptoms include relatively rapid needle desiccation and seedling death, with blackening of the lower stem and root collar inner bark, and dark staining of the underlying wood (fig. 3.3).

Asexual fruiting bodies of these *Diplodia* pathogens are black flask-shaped pycnidia that can be seen with the naked eye or a hand lens. They are produced in dead needles and stems and also are abundant on open female cones (fig. 3.4).

Figure 3.1—*Dead red pine seedlings killed by* Diplodia pinea *in the first season of growth.* Photo by Glen R. Stanosz, University of Wisconsin-Madison.

Figure 3.2—*Distorted red pine shoot killed by* Diplodia pinea *during elongation.* Photo by Glen R. Stanosz, University of Wisconsin-Madison.

Figure 3.3—*Darkly discolored inner bark tissues and stained wood of seedling killed by Diplodia collar rot.* Photo by Glen R. Stanosz, University of Wisconsin-Madison.

Conifer Diseases

3. Diplodia Shoot Blight, Canker, and Collar Rot

They may be solitary or in groups and are often mostly submerged in the host tissue with only short necks erupting through the epidermis and cuticle. Pycnidia are sometimes numerous on dead needle bases (fig. 3.5) below the fascicle sheath. Because numerous fungi produce similar fruiting bodies, spore examination will aid in diagnosis. Conidia produced in the pycnidia are thick-walled, oval, may have one, two, or occasionally more cells, and vary from approximately 30 to 45 by 10 to 15 microns in size (fig. 3.6). These spores are colorless or slightly yellow to light brown when young, becoming very dark brown and opaque with age.

Cultures of *D. pinea* and *D. scrobiculata* can be obtained by placing surface-disinfested, symptomatic needle or stem pieces on malt extract agar or potato dextrose agar amended with lactic acid or streptomycin sulfate to inhibit bacterial growth. A semiselective medium incorporating 0.5 percent w/v (weight:volume) tannic acid in 2 percent water agar facilitates detection of *D. pinea* from asymptomatic seedlings. These fungi will colonize and then produce pycnidia on sterile pine needles placed on the medium surface. Daylight or artificial light will stimulate pycnidia production during incubation at 20 °C to 24 °C (68 °F to 75 °F). Because host ranges and geographic distribution overlap and colony morphology is variable, molecular methods have been developed to differentiate *D. pinea* from *D. scrobiculata*. These methods can be used to identify pathogen isolates or detect their presence on or in host samples without the need to obtain cultures.

Figure 3.4—*Pycnidia of* Diplodia pinea *on scales of an Austrian pine cone.* Photo by Glen R. Stanosz, University of Wisconsin-Madison.

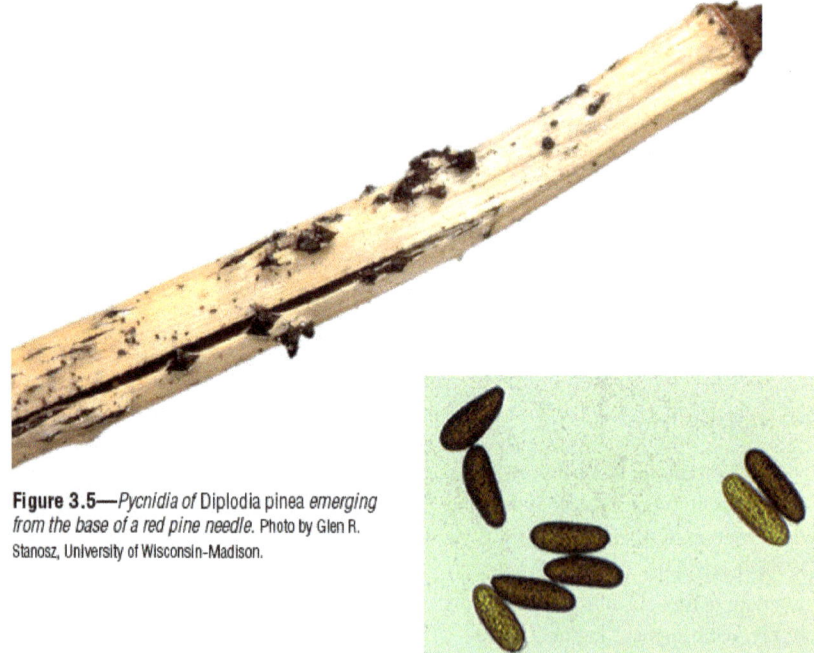

Figure 3.5—*Pycnidia of* Diplodia pinea *emerging from the base of a red pine needle.* Photo by Glen R. Stanosz, University of Wisconsin-Madison.

Figure 3.6—*Conidia of* Diplodia pinea. Photo by Glen R. Stanosz, University of Wisconsin-Madison.

Conifer Diseases

3. Diplodia Shoot Blight, Canker, and Collar Rot

Biology

D. pinea and *D. scrobiculata* survive in and sporulate on dead needles, stems, cones on diseased trees, and debris on the ground. Viable spores can be disseminated by rain splash year round, but are most abundant during spring and early summer when young shoots are most susceptible. Germination occurs rapidly during moist weather, with infection through stomata, directly through the surface of young stems, or through fresh wounds. Pycnidia with conidia can develop within a few weeks after infection, so multiple disease cycles within a single growing season are possible.

Control

Biological

Inherent host resistance is maintained by avoiding both water stress and excessive nitrogen fertilization, which increase susceptibility to disease. Grow nonhost species or less susceptible conifers such as five-needled pines in areas of nurseries where inoculum is present.

Cultural

Eliminate inoculum sources, including host trees in windbreaks and adjacent forests, in the nursery vicinity to minimize disease. Host materials, such as bark, needles, and cones, should not be used as soil amendments or mulches. Practices such as early morning irrigation and decreasing bed densities may promote shoot drying, which reduces infection frequency. Do not move infested seed, diseased seedlings, and seedlings on which the pathogens persist asymptomatically into or out of the nursery.

Chemical

Protectant fungicides can reduce the disease incidence in nursery beds. Repeated applications are required during shoot elongation, however. Note that if inoculum sources are present, fungicide application has not been shown to reduce or eliminate asymptomatic persistence of these pathogens on nursery seedlings that appear to be healthy.

Selected References

Blodgett, J.T.; Bonello, P.; Stanosz, G.R. 2003. An effective medium for isolating *Sphaeropsis sapinea* from asymptomatic pines. Forest Pathology. 33: 395–404.

De Wet, J.; Burgess, T.; Slippers, B.; Preisig, O.; Wingfield, B.D.; Wingfield, M.J. 2003. Multiple gene genealogies and microsatellite markers reflect relationships between morphotypes of *Sphaeropsis sapinea* and distinguish a new species of *Diplodia*. Mycological Research. 107: 557–566.

Peterson, G.W.; Nicholls, T.H. 1989. Diplodia blight. In: Cordell, C.E.; Anderson, R.L.; Hoffard, W.H.; Landis, T.D.; Smith, Jr., R.S.; Toko, H.V., tech. coords. Forest nursery pests. Agriculture Handbook 680. Washington, DC: USDA Forest Service: 31–33.

Smith, D.R.; Stanosz, G.R. 2006. A species-specific PCR assay for detection of *Diplodia pinea* and *D. scrobiculata* in dead red and jack pines with collar rot symptoms. Plant Disease. 90: 307–313.

Stanosz, G.R.; Smith, D.R. 1996. Evaluation of fungicides for control of Sphaeropsis shoot blight of red pine nursery seedlings. Canadian Journal of Forest Research. 26: 492–497.

Conifer Diseases

4. Dothistroma Needle Blight
Glen R. Stanosz

Hosts

Dothistroma needle blight (also known as red band needle blight) damages many pine species and their hybrids and may occasionally affect some other conifers growing in close proximity to diseased pines. Austrian, lodgepole, Monterey, ponderosa, and western white pines can be severely affected, but red and Scots pines appear to be quite resistant. The disease is caused by two fungi that were only recently recognized as distinct species, *Dothistroma septosporum* (sexual stage *Mycosphaerella pini*) and *D. pini* (without a known sexual stage).

Distribution

Dothistroma needle blight occurs widely in the continental United States. Each pathogen's geographic distribution is not completely known, however, with *D. pini* reported primarily from the North Central region.

Damage

Although serious damage to nursery stock is not common, the disease has severely affected landscape trees, Christmas trees, and those in windbreaks, shelterbelts, and some plantations (fig. 4.1). Trees can be almost completely defoliated and may die if disease occurs repeatedly.

Diagnosis

Spots and bands begin to appear on needles early in the fall following infection in the Great Plains region and Central United States, but may develop at any time in areas with more moderate climate. Spots and bands may be tan to orange-red to brown, and needles with bands may die back from their tips (fig. 4.2). Portions of needles between spots and bands often remain green, as may the bases. Lower needles are often affected more severely than those higher in the crowns of established trees. Diseased needles are shed prematurely, but may persist well into the year after infection. On severely affected trees, only the terminal needles may remain, resulting in a "lion's tail" shoot appearance.

The black asexual fruiting bodies of these *Dothistroma* pathogens form within the dead portions of needles and can be seen with the naked eye or a hand lens. Conidia are exuded in a mass through a longitudinal rupture of the epidermis on one or both sides of the fruiting body (fig. 4.3). *D. septosporum* and *D. pini* conidia are colorless, cylindrical, with tapered to rounded tips and slightly

Figure 4.1—*Severe Dothistroma needle blight damage to Monterey pine.* Photo by Glen Stanosz, University of Wisconsin-Madison.

Figure 4.2—*Symptoms of Dothistroma needle blight on lodgepole pine.* Photo from USDA Forest Service Archive at http://www.bugwood.org.

truncated to rounded bases, and usually have two to several cells. *D. septosporum* conidia are 15 to 40 microns long and 2 to 2.5 microns wide. *D. pini* conidia (fig. 4.4) are similar in length, but are 3 to 5 microns wide. Conidia of both *Dothistroma* species can resemble those of the brown spot needle blight pathogen *Lecanosticta acicola*, but *L. acicola* conidia are somewhat olive in color. In the continental United States, *M. septosporum* sexual fruiting bodies have been reported only from the far Western States. These sexual fruiting bodies also rupture the epidermis of necrotic areas of needles and release fusiform, two-celled ascospores that are 10 to 14 microns long and 2.5 to 3.5 microns wide.

Relatively slow growth can make obtaining cultures of these fungi challenging. After surface disinfestation, very small needle segments that bear fruiting bodies can be excised and placed on malt extract agar or potato dextrose agar amended with lactic acid or streptomycin sulfate to inhibit bacterial growth. Alternatively, isolates can be obtained by incubating symptomatic needles in a moist chamber and transferring masses of exuded conidia to culture media. Incubate the conidia at 20 °C (68 °F). Species identity confirmation is facilitated using molecular methods.

Biology

D. septosporum and *D. pini* survive and sporulate in dead areas of needles on diseased trees. In the Great Plains region, one cycle of disease per year has been observed, with spores exuded in moist weather and disseminated by rain splash from spring to fall. Multiple disease cycles within a single growing season may occur in areas with a more moderate climate.

Figure 4.3—*Fruiting bodies of* Dothistroma pini *on Austrian pine needle.* Photo by Glen Stanosz, University of Wisconsin-Madison.

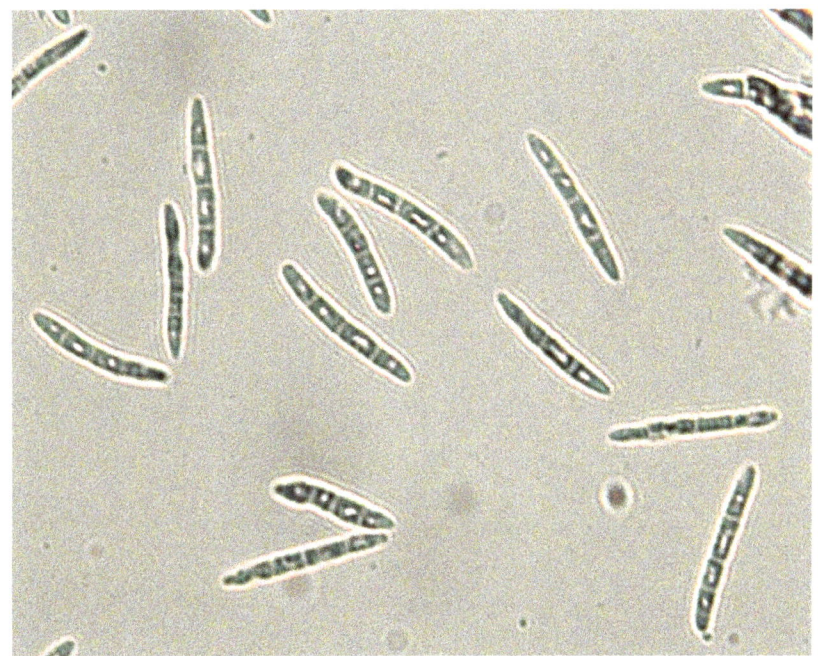

Figure 4.4—*Conidia of* Dothistroma pini. Photo by Glen Stanosz, University of Wisconsin-Madison.

Conifer Diseases

4. Dothistroma Needle Blight

Control

Biological

Variation in resistance within and among pine species has been reported, and nonhosts should be considered in areas where disease is chronic or severe.

Cultural

Elimination of severely affected trees that provide inoculum will decrease incidence and severity of disease in production areas and landscapes. Improving airflow and decreasing wet foliage frequency and duration can also reduce damage. Host species needles should not be used as soil amendments or mulches. Do not move diseased seedlings into or out of the nursery.

Chemical

Protectant fungicides can also reduce disease incidence and severity. Due to the long duration of inoculum availability, however, repeated applications, perhaps throughout the entire growing season, may be required.

Selected References

Barnes, I.; Crouse, P.W.; Wingfield, B.D.; Wingfield, M.J. 2004. Multigene phylogenies reveal that red band needle blight of pines is caused by two distinct species of *Dothistroma*, *Dothistroma septosporum* and *D. pini*. Studies in Mycology. 50: 551–565.

Peterson, G.W. 1989. Dothistroma needle blight. In: Cordell, C.E.; Anderson, R.L.; Hoffard, W.H.; Landis, T.D.; Smith, Jr., R.S.; Toko, H.V., tech. coords. Forest nursery pests. Agriculture Handbook 680. Washington, DC: USDA Forest Service: 34–35.

Sinclair, W.A.; Lyon, H.H. 2005. Diseases of trees and shrubs, 2nd ed. Ithaca, NY: Cornell University Press: 28–30.

Conifer Diseases

5. Eastern and Western Gall Rusts
Michael E. Ostry and Jennifer Juzwik

Revised from chapter by Glenn W. Peterson, William Merrill, and Darroll D. Skilling, 1989.

Hosts

Eastern gall rust (or pine-oak gall rust), caused by one of several formae speciales of the fungus *Cronartium quercuum*, affects jack, Scots, shortleaf, Virginia, and other hard pines. Alternate hosts needed to complete its life cycle include a number of red oak species (section *Lobatae*) and bur oak. The disease may be misidentified as fusiform rust that is common on southern pines. In contrast, western gall rust (or pine-pine gall rust), caused by the fungus *Peridermium harknessii* (syn. *Endocronartium harknessii*), is an autoecious fungus and needs no alternate host. Western gall rust affects many native hard pines of Canada and the United States, including ponderosa, lodgepole, jack, Monterey, Jeffrey, Coulter, knobcone, bishop, and gray pines. Non-native hosts that are naturally infected by the pathogen include Scots, Aleppo, mugo, and Canary Island pines.

Distribution

Eastern gall rust occurs east of the Great Plains from Canada to the Gulf of Mexico where the oak alternate hosts are present. Western gall rust is generally found throughout the pine forests of Western Canada and the United States. In the Eastern United States, it is found as far south as Virginia.

Damage

Seedlings affected by eastern or western gall rust in a nursery usually do not die in the nursery. Losses in the nursery are incurred, however, when diseased seedlings are culled before shipping. Seedling death often occurs after outplanting or in older trees from wind breakage of weakened stems at galls. Reduced wood quality is associated with main stem galls on older trees. Once outplanted, infected seedlings and young trees serve as foci for new infections, especially seedlings infected with *P. harknessii*, because the fungus can spread directly from pine to pine. Infection of oak seedling leaves and older trees by *C. quercuum* results in small necrotic or chlorotic areas, but harm to the tree is minimal. Infected oak, in and around nurseries, serve as foci for new infections of susceptible pine.

Diagnosis

Look for swellings or globose to irregularly shaped galls on the stem (figs. 5.1 and 5.2). Galls are rarely evident until the summer after the year of infection. Galls formed by either western or eastern gall rust fungi are similar in size and shape and the species cannot be distinguished based on gall shape. The presence of the characteristic brown, hair-like spore-bearing structures (telial stage of fungus) on oak leaves in the vicinity, however, suggests that the affected pine seedlings may be infected by the eastern (pine-oak) rust fungus. Conversely, the absence of oak hosts, or the absence of infected leaves on

Figure 5.2—*Aeciospores of* Cronartium quercuum *on gall.* Photo by Michael E. Ostry, USDA Forest Service.

Figure 5.1—*Gall caused by* Cronartium quercuum *on pine seedling.* Photo from USDA Forest Service Archive.

Conifer Diseases

5. Eastern and Western Gall Rusts

nearby oaks suggests that observed galls are caused by the western (pine-pine) rust fungus. The presence of numerous globose to elongate galls on established pine stems and branches in the vicinity of the pine fields or nursery are indicative of the presence of gall rust fungi. These galls are particularly conspicuous in the spring when they are covered by bright orange masses of powdery spores called aeciospores (fig. 5.3). Aeciospores of these rust fungi rarely develop on galls found on nursery seedlings. Although the aeciospores of the two fungi are morphologically identical, they can be distinguished in the laboratory based on characteristics of the germ tube length on culture plates; extended in pine-oak rust and limited in pine-pine gall rust.

Figure 5.3—*Aeciospores of* Cronartium quercuum *being dispersed from galls*. Photo by Michael E. Ostry, USDA Forest Service.

Biology

Aeciospores of *C. quercuum* form on galls of saplings and trees. These spores are wind-disseminated to oak leaves less than 3 weeks old, where they infect and produce repeating spores (urediniospores) and teliospores. Basidiospores produced by the teliospores are then wind-disseminated to pine in late summer-early fall where they infect needles. The fungus grows within needles and into the stem where galls form about 1 year after infection. Timing of the different types of spore production differs by region. *P. harknessii* form orange spores on galls that morphologically resemble aeciospores but function like teliospores. On the galls, *P. harknessii* forms orange spores that morphologically resemble aeciospores but function like teliospores. Following dispersal, these spores germinate and infect needles on current year pine shoots early in the growing season.

Control

Prevention

If practical, remove both oak and infected pines within and around nursery fields. Pines resistant to both gall rusts exist and additional resistant selections may be available in the future. Jack pine seedlings from various sources have been found to be resistant to both gall rusts, but the resistance to both is not correlated. Thus, separate screenings are required for the diseases on this host for areas where both occur.

Cultural

Culling seedlings with galls, either within beds or after lifting, minimizes shipping of gall rust infected stock.

Chemical

Routine fungicide sprays with selected chemicals are effective in preventing infection by gall rust fungi. Fungicide use by nurseries varies by geographic region, however. In the Southeastern United States, fungicide sprays for fusiform rust will be effective against eastern gall rust as well.

Selected References

Anderson, G.W.; French, D.W. 1965. Differentiation of *Cronartium quercuum* and *Cronartium coleosporiodes* on the basis of aeciospore germ tubes. Phytopathology. 55: 171–173.

Burnes, T.A.; Blanchette, R.A.; Stewart, W.K.; Mohn, C.A. 1989. Screening jack pine seedlings for resistance to *Cronartium quercuum* f. sp. *banksianae* and *Endocronartium harknessii*. Canadian Journal of Forest Research. 19: 1642–1644.

Cummins, G.B.; Hiratsuka, Y. 2003. Illustrated genera of rust fungi, 3rd ed. St. Paul, MN: American Phytopathological Society Press. 225 p.

Peterson, G.W.; Merrill, W.; Skilling, D.D. 1989. Eastern and western gall rust. In: Cordell, C.E.; Anderson, R.L.; Hoffard, W.H.; Landis, T.D.; Smith, Jr., R.S.; Toko, H.V., tech. coords. Forest nursery pests. Agriculture Handbook 680. Washington, DC: USDA Forest Service: 36–37.

Sinclair, W.A.; Lyon, H.H. 2005. Diseases of trees and shrubs, 2nd ed. Ithaca, NY: Cornell University Press. 660 p.

White, E.E.; Allen, E.A.; Ying, C.C.; Foord, B.M. 2000. Seedling inoculation distinguishes lodgepole pine families most and least susceptible to gall rust. Canadian Journal of Forest Research. 30: 841–843.

Conifer Diseases

6. Fusiform Rust

Scott A. Enebak and Tom Starkey

Hosts

Fusiform rust, caused by the fungus *Cronartium quercuum* f. sp. *fusiforme*, requires both southern pine and oak trees to complete its life cycle.

Although 32 pine species have been shown to be susceptible to the fungus, the southern pine species most affected by the disease in nurseries are slash and loblolly pine, which are highly susceptible; longleaf and pond pines, which are less susceptible; and pitch and shortleaf pines, which are relatively resistant.

Members of the black oak group are the most common alternate hosts of the fungus, but 33 oak species are susceptible to the disease. Southern red oak and water oak are the most common alternate host for this fungus in the Southern United States.

Distribution

Fusiform rust is indigenous to the Southern United States and can be found from Maryland south to Florida and west to Arkansas and Texas. The disease incidence is highest in a zone approximately 150 miles wide, extending from the South Carolina coast to Texas where environmental conditions favorable for spore production and the susceptible (oak and pine) hosts occur together.

Damage

Fusiform rust is by far the most serious nursery disease of slash and loblolly pine; in contrast, the fungus does not cause any economic loss on oaks. Within the high disease incidence zone, nurseries must take steps to control this disease on pine hosts. Although mortality may not occur in the nursery, infected seedlings rarely survive to age five or may result in poorly formed trees. To minimize losses in the field and disease introductions, infected seedlings must be culled at the nursery before outplanting.

Diagnosis

In the early spring, examine the underside of oak leaves in the vicinity of the nursery for orange urediniospores and brown, hair-like teliospores (fig. 6.1). The presence of urediniospores and telia on oak leaves indicates that if susceptible pine seedlings are in nearby nurseries, they will most likely become infected. Beginning in late summer, examine seedling stems for slight swellings or epidermal discolorations on the main stem above the root collar. On loblolly and slash pine, the typical spindle-shaped stem gall becomes considerably larger and more obvious by the time the seedlings are lifted later in the season (fig. 6.2). On longleaf pine, however, the galls occur right at the groundline and tend to be more globose (fig. 6.3). During the fall, yellow-orange droplets of fluid may be observed on the galls. These droplets contain pycniospores, one of the five spore stages produced by the fungus (fig. 6.4).

Figure 6.1—*Uredinal pustules and hair-like telia of* Cronartium quercuum *f. sp.* fusiforme *on the underside of an oak leaf.* Photo by Robert L. Anderson, USDA Forest Service, at http://www.bugwood.org.

Figure 6.2—*Typical spindle-shaped galls on loblolly pine seedlings infected with fusiform rust.* Photo by Tom Starkey, Auburn University.

Conifer Diseases

6. Fusiform Rust

Figure 6.3—*Typical gall (bottom seedling) formed on longleaf pine seedlings infected with fusiform rust.* Photo by Tom Starkey, Auburn University.

Figure 6.4—*Swelling on seedling stem and pycniospores produced by* Cronartium quercuum *f. sp.* fusiforme. Photo by Scott A. Enebak, Auburn University.

Biology

This disease is caused by an obligate fungal parasite that requires living host tissue for survival. The orange aeciospores produced on pine galls in early spring only infect expanding oak leaves. A few weeks after oak infection, uredinial pustules develop on the underside of oak leaves; the uredinial pustules then produce urediniospores, which may reinfect other young oak leaves. As these infected oak leaves mature, telia will appear and they produce basidiospores. The basidiospores are produced from spring to early summer during periods of high humidity and moderate temperatures, and are the only spores of this fungus that can infect pine needles and shoots. It is at this time of the growing season when seedlings in the nursery are germinating and emerging from the soil and are highly susceptible to basidiospore infection.

Control

Prevention

Use of conifer seed from genetically improved families or selected rust-resistant stock is the most effective and practical long-range approach to lessening the disease's effects on loblolly and slash pine.

Cultural

Only rust-free seedlings should be shipped from the nursery because infected seedlings rarely survive past age five and only serve to introduce or increase the disease in the field. Check seedlings for fusiform rust infection before lifting. If infection is present, then either cull seedlings or destroy infested beds.

Chemical

The use of fungicides provides the most effective control method in forest tree nurseries. Seedling infection can be prevented by applying a registered fungicide as a seed treatment before sowing and then followed with three to four foliar sprays every 2 to 3 weeks from seedling emergence through mid-June. A spreader-sticker can be added to improve coverage and reduce fungicide weathering. The frequency, timing, and coverage of fungicidal sprays are important for effective rust control.

Conifer Diseases

6. Fusiform Rust

Selected References

Anderson, R.L.; McClure, J.P.; Cost, N.; Uhler, R.J. 1986. Estimating fusiform rust losses in five Southeast States. Southern Journal of Applied Forestry. 10: 237–240.

Czabator, F.J. 1971. Fusiform rust of southern pines—a critical review. Res. Pap. SO-65. New Orleans, LA: USDA Forest Service, Southern Forest Experiment Station. 39 p.

Powers, Jr., H.R.; Miller, T.; Belanger, R.P. 1993. Management strategies to reduce losses from fusiform rust. Southern Journal of Applied Forestry. 17: 146–149.

Rowan, S.J. 1984. Bayleton seed treatment combined with foliar spray improves fusiform rust control in nurseries. Southern Journal of Applied Forestry. 8: 51–54.

Rowan, S.J. 1989. Fusiform rust. In: Cordell, C.E.; Anderson, R.L.; Hoffard, W.H.; Landis, T.D.; Smith, Jr., R.S.; Toko, H.V., tech. coords. Forest nursery pests. Agriculture Handbook 680. Washington, DC: USDA Forest Service: 43–44.

Rowan, S.J.; Cordell, C.E.; Affeltranger, C.E. 1980. Fusiform rust losses, control costs, and relative hazard in southern forest tree nurseries. Tree Planters' Notes. 31: 3–8.

Starkey, T.S.; Enebak, S.A. 2010. The use of prothioconazole to control forest nursery diseases of *pinus* spp. In: Proceedings of the 7th meeting of IUFRO Working Party 7.03-04. Report 10-01-01. USDA Forest Service. Southern Region, Forest Health Protection: 92–103.

Conifer Diseases

7. Larch Needle Cast
Katy M. Mallams

Revised from chapter by Sally J. Campbell, 1989.

Hosts

Larch needle cast, caused by the fungus *Meria laricis*, affects many larch species, including western, European, Japanese, hybrid, and Siberian larches.

Distribution

The disease is widespread in North American forests. It is common on nursery stock in the Northwestern United States and Western Canada.

Damage

Severe infection by *M. laricis*, where virtually every needle is killed, will result in seedling mortality. If infection is relatively light, most seedlings will survive; however, seedling growth will be reduced and many seedlings may need to be culled. In the Northwest, good field survival has been obtained for seedlings that were moderately-to-severely infected in the nursery. The most severe damage occurs where larch seedlings are grown in one place for 2 or more years.

Diagnosis

As soon as buds break and needles have expanded, look for yellow to brown spots on needles. Infection usually moves from the needle tip toward the base. The entire needle will turn yellow, then red-brown, sometimes rapidly, and fall prematurely. Needles closer to the ground will be infected first and most heavily (fig. 7.1). Cushion-like tufts of spores emerge through the stomata on the lower surface and occasionally on the upper surface of infected needles. These structures may be visible with a hand lens as white dots, but are more reliably identified by staining the sample and examining it under a microscope (fig. 7.2). The individual spores are transparent, one-celled, peanut-shaped, and 9 to 13 by 3 to 4 microns in size. Fallen needles should also be examined for the fruiting bodies.

Figure 7.1—*Lower needles of bareroot larch seedlings affected by larch needle cast.* Photo by Sally J. Campbell, USDA Forest Service.

Diseased seedlings may occur in small patches or may be scattered throughout the nursery seedbed. Symptoms may become evident very suddenly during or after wet weather. The disease

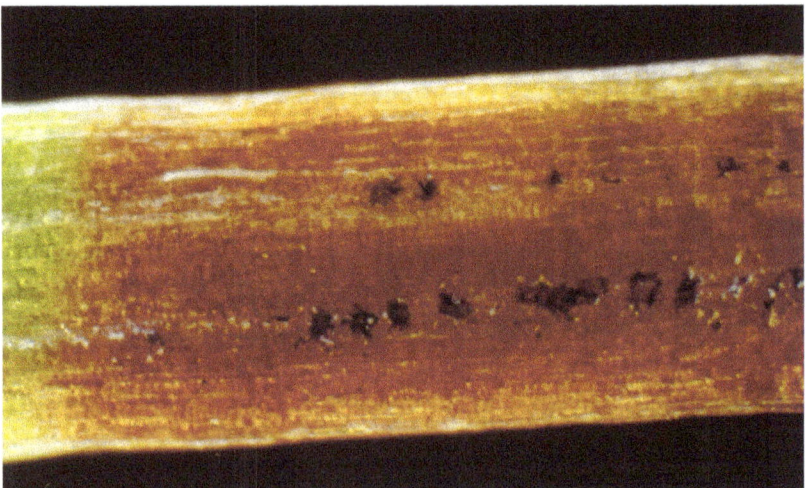

Figure 7.2—*Stained fruiting bodies of* Meria laricis *on a western larch needle.* Photo by Sally J. Campbell, USDA Forest Service.

Conifer Diseases

7. Larch Needle Cast

is sometimes confused with frost damage. In bareroot nurseries, larch needle cast is most noticeable and severe on seedlings 1 year old or older (fig. 7.3); however, disease symptoms may sometimes be observed in the fall of the first growing season. The disease has also been observed in containerized nursery stock (fig. 7.4), where the symptoms appear from middle to late summer.

Biology

The fungus overwinters, presumably as spores or mycelium, on fallen needles or in the dead terminal needles retained by trees or seedlings. Initial infection may occur as soon as buds break and needles have expanded in the early spring. On 1-0 seedlings, infection most likely originates from spores produced on adjacent 2-0 seedlings or on infected larch in the vicinity of the nursery. Infection on 2-0 seedlings originates from infected needles shed the previous winter. Fungal fruiting bodies and conidia are produced after infection; the process of infection and production of conidia continues throughout the spring and summer, provided that weather conditions are favorable. To develop, the fungus needs high humidity and cool to moderate temperatures. Wet spring weather favors disease development. Further infections are halted by hot, dry weather. Infection can also occur in the winter, but symptom development is slow.

Figure 7.4—*Containerized larch seedlings affected by larch needle cast.* Photo ©Her Majesty the Queen in right of Canada, Natural Resources Canada, Canadian Forest Service.

Figure 7.3—*Western larch seedlings severely affected by larch needle cast.* Photo by Sally J. Campbell, USDA Forest Service.

Control

Prevention

To avoid introducing the fungus into disease-free nurseries, grow all larch from seed rather than importing stock from other nurseries. Rotate larch seedbeds so that seedlings are not grown in the same sections for consecutive years and 1-0 seedlings are not grown adjacent to 2-0 seedlings or transplants. To reduce inoculum, remove larch trees adjacent to nurseries; replace larch with other species. Transplant 1-year-old seedlings to a different part of the nursery to avoid reinfecting them with the fungus that overwintered in fallen needles. In nurseries where the disease occurs, reduce irrigation as much as possible and irrigate early in the day to promote rapid drying of foliage. In container nurseries, thoroughly sanitize growing areas between crops. In bareroot nurseries, plowing under diseased needles may reduce the level of inoculum. Since most infection occurs on 1-year-old or older seedlings, outplant 1-year-old seedlings whenever possible.

Chemical

Fungicides have not provided consistent protection against larch needle cast, probably because it is difficult to cover all needle surfaces. To maximize

Conifer Diseases

7. Larch Needle Cast

effectiveness, protective fungicides should be applied at bud swell, a second time 4 weeks later, and subsequently at 2- to 3-week intervals throughout most of the growing season. Additional applications may be needed during extended periods of rain or irrigation. Continue treatments until overhead irrigation is no longer needed or hot, dry weather predominates. Fungicides are usually not needed on 1-0 seedlings unless they are scheduled for 2-0 seedlings.

Selected References

Campbell, S.J. 1989. Larch needle cast. In: Cordell, C.E.; Anderson, R.A.; Hoffard, W.H.; Landis, T.D.; Smith, Jr., R.S.; Toko, H.V., tech. coords. Forest nursery pests. Agriculture Handbook 680. Washington, DC: USDA Forest Service: 47–48.

Hubert, E.E. 1954. Needle cast diseases of western larch. Bull. 215. Moscow, ID: University of Idaho Agricultural Experiment Station and Cooperative Extension Service. 6 p.

Peace, T. 1962. Pathology of trees and shrubs. Oxford, England: Clarendon Press. 753 p.

Sinclair, W.A.; Lyon, H.H.; Johnson, W.T. 1996. Diseases of trees and shrubs. Ithaca, NY: Cornell University Press. 575 p.

Thomson, A.; Dennis, J.; Trotter, D.; Shaykewich, D.; Banfield, R. 2009. Diseases and insects in British Columbia forest nurseries. http://forestry-dev.org/diseases/nursery/pests/mlaricis_e.html. Accessed October 2010.

Conifer Diseases

8. Needle Cast Diseases of Pines
Katy M. Mallams

Hosts

Nearly all pine species are susceptible to infection by one or more of the fungi that cause needle cast diseases. Most of the needle cast fungi are weak pathogens and have specific host preferences. However, some host-pathogen combinations can result in significant outbreaks of disease when weather and site conditions are favorable. Nonnative species such as Scots, Austrian, and Japanese black pine are especially susceptible to needle cast diseases and can suffer severe damage in the nursery. With the exception of red pine, seedlings of native pine species are normally not seriously affected in the nursery. Three species of fungi, *Cyclaneusma minus*, *Lophodermium seditiosum*, and *Ploioderma lethale*, are the most common cause of severe needle cast disease outbreaks in nurseries that grow pines. These pathogens and their host species are listed in table 8.1.

Distribution

The fungi that cause pine needle cast diseases are widely distributed throughout North America and can usually be found wherever their hosts are present.

Damage

The cohort of needles affected depends on which fungus is involved. *L. seditiosum* primarily infects newly elongated needles on the current season's shoots. *P. lethale* mainly infects 2-year-old needles. *C. minus* infects all needles except those of the current year. Damage usually becomes evident the year following infection when the affected needles die and fall prematurely. Needle cast diseases seldom result in mortality, but extensive needle loss may result in reduced growth and loss of vigor. Severely discolored or defoliated seedlings may lose their sale value because of poor appearance. If infected, outplanted seedlings may perform poorly and may serve as a source of inoculum for infection of other nearby pines. For these reasons, diseased seedlings should be culled.

Diagnosis

Lophodermium and *Ploioderma* needle cast symptoms become visible on current-year needles between late autumn and the next spring. Initially, yellow to reddish-brown spots develop on infected needles (fig. 8.1). These spots can be confused with damage caused by sucking insects. As the spots enlarge, needles infected by *L. seditiosum* take on a mottled appearance. By summer, they turn brown and fall from the tree. Needles infected by *P. lethale* turn grayish-brown but remain green near the base. Seedlings severely affected by either disease may appear scorched (fig. 8.2). Fruiting bodies (apothecia) of *L. seditiosum* and *P. lethale* appear beginning in late spring on dead portions of infected needles and on fallen needles. The fruiting bodies of *L. seditiosum* are small, black football-shaped structures visible with the naked eye (fig. 8.3). Fruiting bodies of *P. lethale* appear as elongated black lines (fig. 8.4).

Cyclaneusma needle cast symptoms appear on 2-year-old needles in late summer or autumn the year after the needles were infected. Spots on infected needles are initially light green. The spots enlarge into yellow bands, and eventually the entire needle becomes yellow, often with transverse brown bands. Fruiting bodies of *C. minus* initially develop in the brown areas, and later on the entire lower surface of the infected needles. These fruiting bodies are initially small and inconspicuous, gradually elongating.

Table 8.1—*Needle cast fungi causing disease on pine seedlings and their hosts.*

Species	Pines highly susceptible to damage in nurseries	Other recorded pine hosts, not commonly damaged in nurseries
Cyclaneusma minus	Scots, Austrian, and Monterey	Cuban, gray, Jeffrey, limber, lodgepole, mugo, ponderosa, Virginia, eastern white, and others
Lophodermium seditiosum	Austrian, red, Scots[1], and Monterey[2]	Aleppo, rough-barked Mexican, and Virginia
Ploioderma lethale	Austrian, red, and Japanese black	Austrian, Japanese black, Cuban, loblolly, pitch, pond, red, sand, shortleaf, slash, spruce, Table Mountain, and Virginia

[1] Particularly short-needle strains originating in France and Spain.
[2] When grown outside its native range.

Figure 8.1—*Typical needle spots on red pine caused by* Lophodermium *species.* Photo by USDA Forest Service, North Central Research Station Archive, at http://www.bugwood.org.

Conifer Diseases

8. Needle Cast Diseases of Pines

Figure 8.2—*From a distance, diseased pine foliage appears scorched.* Photo by Joseph O'Brien, USDA Forest Service, at http://www.bugwood.org.

The fungal tissue erupts through the epidermis, which typically splits in two thin flaps to expose the fungal tissue. The fungal tissue is yellow to orange, later becoming white or tan as the spore-producing surface is exposed (fig. 8.5).

Identifying the species causing needle cast diseases often requires microscopic examination by a specialist. Many saprophytic or weakly pathogenic fungi that grow on dead or dying needles but do not cause disease have fruiting bodies that may appear very similar to the pathogenic species. Accurate identification is necessary to determine if the symptoms are due to a needle cast disease and the species involved. Accurate identification is also necessary to time chemical treatments for effective control.

Biology

All three species release spores when mature fruiting bodies are moistened by rain or irrigation, although the seasonality of spore release differs. *C. minus* spores are released throughout the year when temperatures are above freezing, peaking 4 to 6 hours after rain begins. *L. seditiosum* spores are released in late summer and fall, and occasionally at other times of the year if conditions are favorable. *P. lethale* spores are released during late spring and early summer. These fungi also require moist conditions for spore germination and infection. *L. seditiosum* and *P. lethale* spores infect only current-year needles. *C. minus* spores infect current-year needles in late spring and older needles from spring through autumn. Year-to-year variations in weather and differences in climate among geographic areas will affect the exact timing of spore release and infection by these fungi.

Spores may be produced on infected trees in the vicinity of the nursery, on infected nursery stock, and on fallen needles. Infected seedlings transplanted from other facilities and pine needles used as mulch are potential sources of inoculum. *L. seditiosum* also fruits on cone scales and can be introduced into nurseries on cone fragments mixed with seeds. Severe outbreaks of needle cast diseases are most likely to occur after cool, moist weather in spring, summer, or early fall creates favorable conditions for buildup of abundant populations of spores and widespread infection of susceptible hosts.

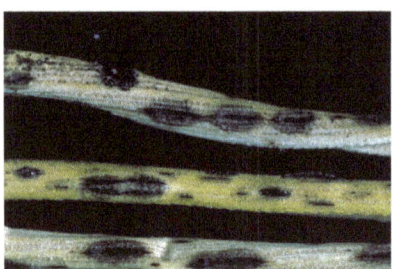

Figure 8.3—*Fruiting bodies of* Lophodermium *species on pine needles.* Photo by Robert L. Anderson, USDA Forest Service, at http://www.bugwood.org.

Figure 8.4—*Fruiting bodies of* Ploioderma lethale *on pine needles.* Photo by Department of Plant Pathology and Plant-Microbe Biology, Cornell University.

Conifer Diseases

8. Needle Cast Diseases of Pines

Figure 8.5—*Fruiting bodies of* Cyclaneusma minus *on Scots pine needles.* Photo by Joseph O'Brien, USDA Forest Service, at http://www.bugwood.org.

Control

Prevention

Ideally, nurseries that grow pines should be located as far as possible from pine forests. Pines in windbreaks and residential areas adjacent to nurseries can also be sources of inoculum. Seed should be inspected and thoroughly cleaned before sowing to remove cone scales that may harbor inoculum. Pine seedlings received from other nurseries should be inspected carefully before transplanting. Do not use pine needle mulch.

Cultural

Allow needles to remain wet for the shortest possible time. Low seedling density will encourage good air circulation, allowing seedlings to dry as rapidly as possible after rain and irrigation. Drip irrigation is ideal for keeping foliage dry. If overhead sprinklers are used, watering in the morning will help seedlings dry faster. After pine seedlings are lifted, fallen needles should be removed or tilled in if susceptible pines will be grown nearby the following year.

Chemical

Fungicides can be used to protect healthy foliage from infection. They must be applied just before and during spore release periods, and before infection occurs. To be effective, thorough coverage of the foliage is required. The timing of fungicide applications should be based on proper identification of the causal fungus, local weather conditions, and the product label recommendations. Wet years will require more frequent applications. High-density plantings and areas with a history of needle cast diseases may also require more frequent fungicide treatments.

Selected References

Jones, R.K.; Benson, D.M. 2001. Diseases of woody ornamentals and trees in nurseries. St. Paul, MN: The American Phytopathological Society Press. 482 p.

Minter, D.W. 1981. *Lophodermium* on pines. Mycological Papers. 147: 1–54.

Minter, D.W.; Millar, C.S. 1980. Ecology and biology of three *Lophodermium* species on secondary needles of *Pinus sylvestris*. European Journal of Forest Pathology. 10: 169–181.

Moorman, G.W. 2006 (December). Needle cast diseases. Plant disease facts. Cooperative Extension, The Pennsylvania State University, Department of Plant Pathology. http://www.ppath.cas.psu.edu/extension/plant_disease/pdf%20Woody/needlecasts.pdf. Accessed October 2010.

Nicholls, T.H.; Brown, H.D. 1975. How to identify *Lophodermium* and brown spot diseases on pine. St. Paul, MN: USDA Forest Service, North Central Forest Experiment Station. 5 p.

Nicholls, T.H.; Skilling, D.D. 1974. Control of Lophodermium needle cast in forest nurseries and Christmas tree plantations. Res. Pap. NC-llO. St. Paul, MN: USDA Forest Service, North Central Forest Experiment Station. 11 p.

Sinclair, W.A.; Lyon, H.H.; Johnson, W.T. 1996. Diseases of trees and shrubs. Ithaca, NY: Cornell University Press. 575 p.

Staley, J.M.; Nicholls, T.H. 1989. Lophodermium needle cast. In: Cordell, C.E.; Anderson, R.A.; Hoffard, W.H.; Landis, T.D.; Smith, Jr., R.S.; Toko, H.V., tech. coords. Forest nursery pests. Agriculture Handbook 680. Washington, DC: USDA Forest Service: 49–51.

Stone, J. 1997. Needle blights and needle casts. In: Hansen, E.M.; Lewis, K.J. Compendium of conifer diseases. St. Paul, MN: The American Phytopathological Society Press: 53–54.

Conifer Diseases

9. Passalora Blight
Charles S. Hodges and Michelle M. Cram

Hosts

Passalora blight (Cercospora blight), caused by *Passalora sequoiae* (syn. *Cercospora sequoiae*), is associated with several members of the cypress family (*Juniperus, Cupressus, Chamaecyparis, Thuja, Cryptomeria*) and the bald cypress family (*Taxodium, Sequoia, Sequoiadendron*). Nursery seedling infection by *P. sequoiae* has been reported on eastern redcedar, giant sequoia, cypress, and bald cypress. A similar disease on cypress and junipers, also commonly known as Passalora blight, is caused by *Pseudocercospora juniperi* (syn. *Cercospora sequoiae* var. *juniperi*).

Distribution

Both fungi that produce Passalora blight have wide distribution in the Southern and Midwestern United States. *P. sequoiae* also has been reported from the West Coast and Hawaii.

Damage

Damage to seedlings in the nursery by both fungi varies from very light to almost complete defoliation. In southern forest nurseries, *P. sequoiae* has caused serious damage to eastern redcedar; while in the North Central States, *P. juniperi* has been a more significant problem on Rocky Mountain juniper in windbreaks and other plantings. No information is available on infected seedling survival and growth rates after outplanting; however, Passalora blight is likely to continue to develop in the field on seedlings infected in the nursery. The danger of introducing the fungus into new areas exists if infected seedlings are outplanted.

Diagnosis

Seedling foliage with Passalora blight becomes brown along the stem and on the lower branches. The disease continues to develop upward and outward until only the upper branch tips remain green on severely infected seedlings (fig. 9.1). This damage pattern differentiates Passalora blight from the more common Phomopsis blight on juniper (see chapter 12) where the disease develops from the tips of the branches inward.

The fruiting structures of both fungi are very similar. Dark brown pustule-like structures, or stromata, develop on needles shortly after they turn brown, and are easily visible with a hand lens. The stromata are 50 to 115 microns in diameter for *P. sequoiae* and 60 to 200 microns in diameter for *P. juniperi*. Yellow-brown to brown conidiophores protrude from the stromata, forming a compact layer over the surface. The conidiophores are geniculate (bent abruptly; knee shape) and are 50 to 125 by 4 to 6 microns for *P. sequoiae* and 20 to 45 by 3 to 5 microns for *P. juniperi* (fig. 9. 2). The two fungi can be distinguished by their conidial characteristics. *P. sequoiae* spores are yellow-brown, cylindrical, slightly tapering, mostly 5 to 6 septate, echinulate (prickles) and average 40.5 by 5.4 microns (fig. 9.3). *P. juniperi* spores are olive-brown, cylindrical, mostly 5 to 6 septate, slightly echinulate, and average 40.8 by 3.1 microns.

Figure 9.1—*Passalora blight symptoms on eastern redcedar.* Photo courtesy of the University of Wisconsin.

Conifer Diseases

9. Passalora Blight

Figure 9.2—*Conidiophores of* Passalora sequoiae. Photo by Charles S. Hodges.

Biology

Initial infection of first-year nursery seedlings usually comes from older infected plants in the nursery or from infected nearby windbreaks or landscape plantings of susceptible hosts. Spring inoculum is produced from fungal structures that overwintered on needles of infected trees or seedlings. Disease symptoms develop within 2 to 3 weeks, and fruiting bodies form after the foliage turns brown. The resulting conidia are spread primarily by wind. Production of conidia and new infections can occur throughout the spring and summer. Wet weather and moderate temperatures favor disease development.

Control

Prevention

Do not use known hosts of the two fungi as windbreak or landscape plants in or near the nursery. If known-host plants are present, remove any that are infected. Variation in resistance to *P. juniperi* among genotypes of eastern redcedar and Rocky Mountain juniper has been reported. In general, eastern redcedar has been found to be more resistant to *P. juniperi* than Rocky Mountain juniper. If available, use seed sources resistant to Passalora blight.

Cultural

Use irrigation early in the morning to promote rapid drying of foliage. After a disease outbreak or lifting, incorporate residual seedlings and debris into the soil to reduce spring inoculum.

Chemical

Fungicides labeled for use in forest nurseries against leaf or needle diseases on conifers can be used to reduce infection and spread of Passalora blight. A standard spray schedule may be necessary throughout the growing season to control this disease in some areas.

Selected References

Crous, P.W.; Braun, U. 2003. *Mycosphaerella* and its anamorphs: names published in *Cercospora* and *Passalora*. Utrecht, The Netherlands: Centraalbureau voor Schimmelcultures. 571 p.

Hodges, C.S. 1962. Comparison of four similar fungi from *Juniperus* and related conifers. Mycologia. 54: 62–69.

Hodges, Jr., C.S.; Peterson, G.W. 1989. Cercospora blight of Juniper. In: Cordell, C.E.; Anderson, R.A.; Hoffard, W.H.; Landis, T.D.; Smith, Jr., R.S.; Toko, H.V., tech. coords. Forest nursery pests. Agriculture Handbook 680. Washington, DC: USDA Forest Service: 29–30.

Peterson, G.W. 1977. Epidemiology and control of a blight of *Juniperus virginiana* caused by *Cercospora sequoiae* var. *juniperi*. Phytopathology. 67: 234–238.

Zhang, J.W.; Klopfenstein, N.B.; Peterson, G.W. 1997. Genetic variation in disease resistance of *Juniperus virginiana* and *J. scopulorum* grown in eastern Nebraska. Silvae Genetica. 46: 1–16.

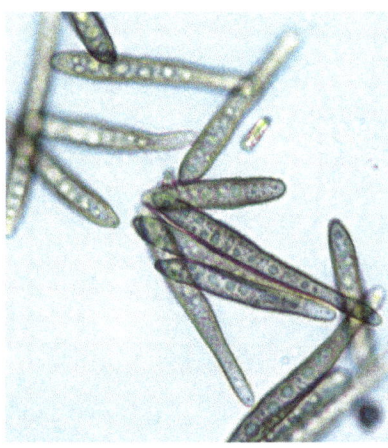

Figure 9.3—*Conidia of* Passalora sequoiae. Photo by Charles S. Hodges.

Conifer Diseases

10. Pestalotiopsis Foliage Blight
Scott A. Enebak

Revised from chapter by Charles E. Affeltranger and Charles E. Cordell, 1989.

Hosts

Pestalotiopsis foliage blight, caused by *Pestalotiopsis funerea* (syn. *Pestalotia funerea*), affects seedlings of many conifer hosts. Eastern white pine has been particularly sensitive to the pathogen in southern forest nurseries.

Distribution

Pestalotiopsis foliage blight occurs in forest tree nurseries throughout the United States wherever susceptible conifer hosts are grown.

Damage

Widespread mortality seldom occurs in the nursery; however, the fungus has been associated with damping-off, root and collar rot, shoot or tip blight, twig dieback, and stem cankers. Outplanting of infected and severely damaged seedlings (defoliation, tip blight, etc.) can result in greater mortality and reduced growth when compared with apparently healthy seedlings.

Diagnosis

The first symptoms appear from late August to October. Infected seedlings have small yellowish spots on the needles that eventually coalesce and turn brown. Needle browning will progress rapidly toward the needle tips. Symptoms appear in small patches throughout the nursery (fig. 10.1) and, if conditions are favorable, seedlings in the entire nursery bed may look brown and scorched (fig. 10.2). Extensive defoliation can occur within a few weeks. By the time seedlings are harvested (December to March), defoliation can be so intense that only a few green needles remain near the terminal bud. On affected needles, small, shiny, black, fruiting bodies can be seen with the naked eye or with a 10x hand lens. Under moist conditions, long, black ribbons of spores are exuded from the fruiting bodies. Individual spores are mostly five-celled and are 22 to 32 by 7 to 13 microns. The spores are ornamented with two to three slender appendages on one end and a single appendage on the other (fig. 10.3).

Figure 10.1—*Small infection center of Pestalotiopsis blight on eastern white pine.* Photo by Scott A. Enebak, Auburn University.

Biology

P. funerea is known as both a pathogen and an opportunistic colonizer of stressed and damaged conifers. The fungus produces acervuli fruiting bodies on browned needles and stems. Conidia erupt from the fruiting bodies when moistened in a sticky black matrix or hornlike projections called tendrils. Rain splash, irrigation, and perhaps insects spread the spores. Infection is correlated with extended periods of above-average rainfall during the growing season and there appears to be a moisture relationship as disease increases with increasing precipitation. The disease incidence is highest and damage most severe in densely stocked seedbeds.

Figure 10.2—*Widespread defoliation of eastern white pine due to Pestalotiopsis blight.* Photo by Scott A. Enebak, Auburn University.

Conifer Diseases

10. Pestalotiopsis Foliage Blight

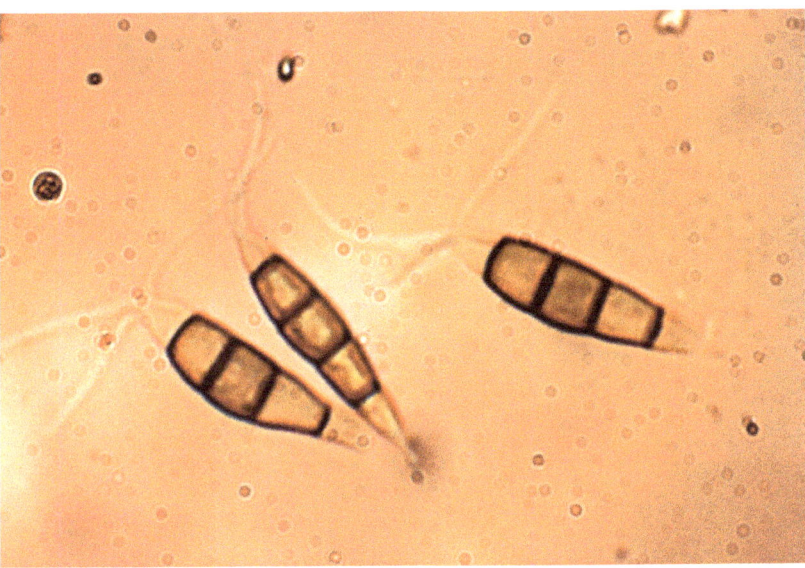

Figure 10.3—*Conidia of* Pestalotiopsis *species associated with foliage blight of conifers.* Photo by Michelle M. Cram, USDA Forest Service.

Control

Prevention

Use of disease-free mulches such as hydro-mulch, sawdust, and pine bark is recommended. Avoid pine straw mulch, since it may carry the fungus into the nursery. In nurseries that grow white pine seedlings, avoid using white pine for windbreaks, which can be an inoculum source. Cull infected seedlings and never ship diseased seedlings between nurseries.

Cultural

Plant at densities of fewer than 270 seedlings per m^2 (25 seedlings per ft^2) as high seedling densities increase moisture retention of the foliage and decrease air movement—conditions that favor the spread of the fungus. Irrigate during the early morning hours, when seedlings dry most quickly. Remove and destroy seedlings that have 50 percent or more of their foliage discolored or that are 25 percent or more defoliated. More intensive practices may be needed when culling white pine seedlings that may be used for Christmas tree stock.

Chemical

Fungicides registered for use in conifer nurseries have been effective in controlling this disease. If the disease is prevalent, apply foliar sprays to seedlings from May to October. Use additional fungicides during periods of excessive rainfall. Soil fumigation has been shown to be effective in reducing the disease incidence in problem areas.

Selected References

Affeltranger, C.E.; Cordell, C.E. 1989. Eastern White Pine Foliage Blight. In: Cordell, C.E.; Anderson, R.A.; Hoffard, W.H.; Landis, T.D.; Smith, Jr., R.S.; Toko, H.V., tech. coords. Forest nursery pests. Agriculture Handbook 680. Washington, DC: USDA Forest Service: 38–39.

Graham, J.E.; Nichols, C.R.; Affeltranger, C.E. 1974. An evaluation of five fungicides for control of a white pine foliage disease at the Piedmont Forest Tree Nursery in South Carolina. Asheville, NC: USDA Forest Service, State and Private Forestry, Forest Pest Management. 7 p.

Nag Raj, T.R. 1993. Coelomycetous anamorphs with appendage-bearing conidia. Mycologue Publications. Waterloo, Ontario, Canada. 1,101 p.

Vermillion, M.T. 1950. A needle blight of pine. Lloydia. 13: 196–197.

Conifer Diseases

11. Phoma Blight
Robert L. James

Revised from chapter by Michael D. Srago, Robert L. James, and John T. Kliejunas, 1989.

Hosts

Phoma blight is associated primarily with *Phoma eupyrena*, although a few other *Phoma* species are sometimes isolated from diseased plant tissues. The disease primarily affects Douglas-fir; red and white fir; mugo, lodgepole, and ponderosa pines; and Engelmann spruce.

Distribution

Phoma blight probably occurs at low levels within bareroot nurseries in most Western States.

Damage

Phoma blight causes chlorotic and necrotic foliage, tip dieback, defoliation, and mortality of 1-0 bareroot seedlings. Stem cankers of 2-0 stock may also occur. Losses vary, but can be significant during certain years at some nurseries.

Diagnosis

On young 1-0 bareroot seedlings, chlorotic needles become evident near the groundline. Foliage covered with soil becomes necrotic. As the disease progresses, the entire seedling becomes chlorotic and finally dies. On Douglas-fir seedlings, infected needles often turn a golden brown and frequently drop prematurely. Terminal and lateral branch dieback or blight occurs on both Douglas-fir and true firs but is more common on the latter. Dieback starts at or near the buds, progresses down the stem, and, if continued, results in seedling death (figs. 11.1 and 11.2). Stem cankers associated with colonization by *P. eupyrena* may occur on older (2-0) Douglas-fir seedlings, but mortality is rare on older seedlings. *Phoma* produces black fruiting bodies (pycnidia) on dead needles, stems, and canker margins. Other fungi (*Sphaeropsis, Sirococcus*), however, may form similar looking fruiting bodies on diseased conifer seedlings. Microscopic examination of spores from fruiting bodies is necessary to verify which pathogen is involved. *Phoma* spores are hyaline and one-celled (fig. 11.3). Size varies according to species; spores of *P. eupyrena* are 3 to 6 by 1.5 to 3 microns.

Figure 11.1—*Blight in terminal bud of a true fir seedling. Small black fruiting bodies of the fungus are visible on dead needles.* Photo by Robert L. James, USDA Forest Service.

Figure 11.2—*Dieback progressing down the stem of a true fir seedling. Note discoloration on stem and at base of needles.* Photo by Robert L. James, USDA Forest Service.

Biology

Phoma species are common soil inhabitants. Overhead irrigation or rain splash may result in excessive soil collar buildup around young seedling stems. *Phoma* can invade seedlings from soil collars, usually through the lower needles. *Phoma* then spreads up the seedling crown, killing needles until the seedling is defoliated. *Phoma* also frequently kills new buds. On older seedlings, soilborne *Phoma* spores can infect needles;

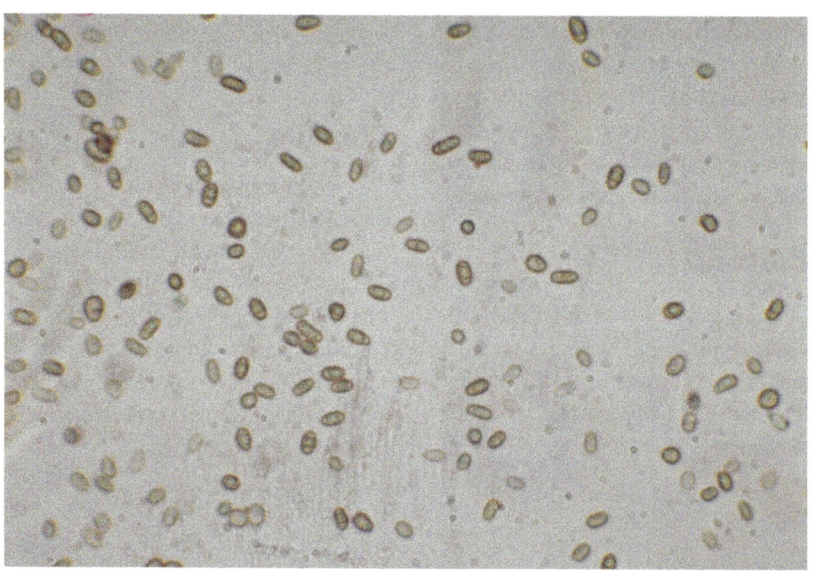

Figure 11.3—*Clear, oval, one-celled spores of* Phoma *species.* Photo by Robert L. James, USDA Forest Service.

Conifer Diseases

11. Phoma Blight

colonization then progresses to stems, eliciting cankers. Seedlings weakened by nutrient imbalances, such as excess calcium and iron, are especially susceptible to attack.

Control

Cultural

Sowing bareroot beds early to increase seedling height during the first year may be helpful in reducing damage and mortality. Foliage above soil cones formed during the winter following the first growing season seldom becomes infected. Mulches that reduce soil cone formation help limit Phoma blight incidence.

Chemical

Fumigating soil before planting reduces or eliminates potentially pathogenic fungi in nurseries, including *Phoma*. Most *Phoma* species are good saprophytes, and high soil populations may build up on incorporated cover crops and organic amendments. When disease symptoms become noticeable, fungicide applications at 2- to 4-week intervals during the dormant season (October to April) reduces losses.

Selected References

James, R.L. 1979. Lodgepole pine seedling chlorosis and mortality at the Bessey Nursery, Nebraska. Bio. Eval. R2-79-2. Denver, CO: USDA Forest Service, State and Private Forestry, Rocky Mountain Region. 10 p.

James, R.L. 1983. Phoma blight of conifer seedlings. Nursery Disease Notes 3. USDA Forest Service, Northern Region, Forest Pest Management. 8 p.

James, R.L. 1984. Characteristics of *Phoma* isolates from nurseries of the Pacific Northwest Region. Nursery Disease Notes 14. USDA Forest Service, Northern Region, Forest Pest Management. 26 p.

James, R.L. 1989. Effects of fumigation on soil pathogens and beneficial microorganisms. In: Landis, T.D., tech. coord. Proceedings: Intermountain Forest Nursery Association Meeting. GTR-RM-184. USDA Forest Service, Rocky Mountain Research Station. 29–34.

James, R.L.; Hamm, P.B. 1985. Chlamydospore-producing species of *Phoma* from conifer seedlings in Pacific Northwest forest tree nurseries. Proceedings of the Montana Academy of Sciences. 45: 26–36.

Kliejunas, J.T.; Allison, J.R.; McCain, A.H.; Smith, R.S. 1985. Phoma blight of fir and Douglas-fir seedlings in a California nursery. Plant Disease. 69: 773–775.

Srago, M.D. 1978. Nursery disease problems: Phoma blight. In: Gustafson, R.W., ed. Proceedings, 1978 nurseryman's conference and seed processing workshop; 1978 August 7-11; Eureka, CA. San Francisco, CA: USDA Forest Service, Pacific Southwest Region, State and Private Forestry: B–138.

Srago, M.D.; James, R.L.; Kliejunas, J.T. 1989. Phoma blight. In: Cordell, C.E.; Anderson, R.L.; Hoffard, W.H.; Landis, T.D.; Smith, Jr., R.S.; Toko, H.V., tech. coords. Forest nursery pests. Agriculture Handbook 680. Washington, DC: USDA Forest Service: 54–55.

Conifer Diseases

12. Phomopsis Blight
Edward L. Barnard

Hosts

Many members of the Cupressaceae (cypress family) are susceptible to infection by *Phomopsis juniperovora*. Juniper species are particularly susceptible, followed by cypress and cedar species. The pathogen is occasionally detected on other gymnosperm hosts, but these occurrences are of relatively little importance.

Distribution

P. juniperovora and the blight it causes occur throughout the eastern half of the United States, California, and the Pacific Northwest.

Damage

Phomopsis blight can cause severe seedling loss in nursery beds leading to the failure of an entire crop (fig. 12.1). This disease will initially appear as a tip or shoot blight and occur in individual patches of the nursery beds. As the blight spreads to adjacent seedlings, the initially infected seedlings may progress from dieback to mortality. Seedlings affected by Phomopsis blight perform poorly and have high mortality rates following outplanting.

Diagnosis

P. juniperovora infects only immature, succulent foliage; therefore, infection and associated symptoms are initiated at growing seedling terminals and branch tips (fig. 12.2). Infections may appear initially as small yellow spots on young foliage but rapidly progress to a blight resulting in reddish browning and eventually graying foliage with necrosis of small twigs and stems. Infected stem and branch tips often curl into a

Figure 12.1—*Seedbed of eastern redcedar badly damaged by* Phomopsis juniperovora. Photo by Gregory A. Hoss, Missouri Department of Conservation.

Figure 12.2—*Tips of eastern redcedar seedlings showing symptoms of Phomopsis blight.* Photo by Edward L. Barnard, Florida Division of Forestry.

Conifer Diseases

12. Phomopsis Blight

"shepherd's crook" as infected tissues dry and become gray. Small, pale orange, grayish or black, pimple-like pycnidia (asexual spore-producing structures) develop in, and erupt from, necrotic tissues. Under high moisture conditions (humidity, rainfall, or irrigation), active pycnidia exude pale yellowish to cream-colored masses or hair-like tendrils of conidia (fig. 12.3). Two different conidia spores are produced by pycnidia (fig. 12.4). Alpha-spores are elliptical, colorless, single spores with two oil drops (7.5 to 10.0 by 2.2 to 2.8 microns). Beta-spores are needlelike, colorless, single spores that are curved at one end (20 to 27 by 1 microns). The alpha-spores are the only conidia that germinate and sometimes are the only spores produced by pycnidia.

Other foliage blights affecting members of the Cupressaceae can sometimes be confused with Phomopsis blight. Infections caused by *Passalora sequoiae* (equal to *Cercospora sequoiae* and *Asperosporium sequoiae*) typically progress from older needles or scales on lower branches, spreading upward and outward over time. Infections caused by *Kabatina juniperi* and *Sclerophoma pythiophyla* are similar in development and appearance to those caused by *P. juniperovora* and are best distinguished by microscopic examination.

Biology

P. juniperovora overwinters in host tissues infected the previous year. Conidia (asexual spores) are produced in the spring in pycnidia formed the previous year. These spores are released and disseminated by splashing rain or overhead irrigation. New infections occur on succulent young foliage tissues with new pycnidia and sporulation developing on infected tissues within 3 to 4 weeks.

Control

Cultural

Avoid excessive irrigation and high seedbed densities as these conditions favor moisture buildup and retention and disease development. Do not sow seed of susceptible hosts adjacent to beds containing older host seedlings and avoid using junipers and other hosts as nursery windbreaks, because older seedlings and adjacent trees may be problematic inoculum sources. Infected seedlings should be rogued from seedbeds to minimize spread of infective spores. Irrigation should be applied early during the day to enable quick drying of the foliage. In some areas and for some uses, disease resistant junipers may be a helpful alternative management strategy.

Chemical

Protective fungicides applied at 7- to 10-day intervals during seasons of active seedling growth are helpful.

Selected References

Hodges, C.S.; Green, H.J. 1961. Survival in the plantation of eastern redcedar seedlings infected with *Phomopsis juniperovora* in the nursery. Plant Disease Reporter. 45: 134–136.

Peterson, G.W.; Hodges, Jr., C.S. 1989. Phomopsis blight of junipers. In: Cordell, C.E.; Anderson, R.A.; Hoffard, W.H.; Landis, T.D.; Smith, Jr., R.S.; Toko, H.V., tech. coords. Forest nursery pests. Agriculture Handbook 680. Washington, DC: USDA Forest Service: 56–58.

Sinclair, W.A.; Lyon, H.H. 2005. Diseases of trees and shrubs, 2nd ed. Ithaca, NY: Cornell University Press. 660 p.

Figure 12.3—*Pycnidia of* Phomopsis juniperovora *on symptomatic foliage of eastern redcedar exuding string-like tendrils of asexual spores (conidia).* Photo by Edward L. Barnard, Florida Division of Forestry.

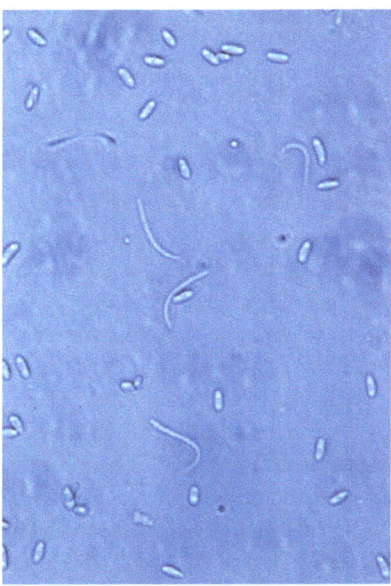

Figure 12.4—*Alpha-spores (elliptical) and beta-spores (needlelike) of* Phomopsis juniperovora. Photo by Michelle M. Cram, USDA Forest Service.

Conifer Diseases

13. Phomopsis Canker
Katy M. Mallams

Hosts

Phomopsis canker, caused by the fungus *Phomopsis lokoyae* (sexual state: *Diaporthe lokoyae*), affects coastal and Rocky Mountain Douglas-fir. A similar species, *P. occulta,* causes cankers on Sitka spruce, western hemlock, western larch, and western redcedar.

Distribution

Phomopsis canker is widespread in the Pacific Northwest and northern California.

Damage

Phomopsis canker occurs sporadically in nurseries and plantations, especially 1 to 2 years after droughts and damaging frost events. It is most common on seedlings in the second growing season, although 1-year-old seedlings are occasionally affected. *Phomopsis* species cause cankers at the base of new growth. Shoots above the canker are killed, but lateral shoots below the canker are not affected and normally will develop into a new leader. The greatest impact of the disease in the nursery is a reduction in seedling quality and an increase in the number of culls. In plantations, the disease causes top-kill and occasional mortality of small trees. Mortality is infrequent and is usually scattered throughout nursery beds. Widespread damage has occurred occasionally in nurseries when bud break coincided with prolonged wet weather.

Diagnosis

Cankers become visible on stems and branches of current-year growth in spring and summer (fig. 13.1). Cankers are usually sharply defined and appear

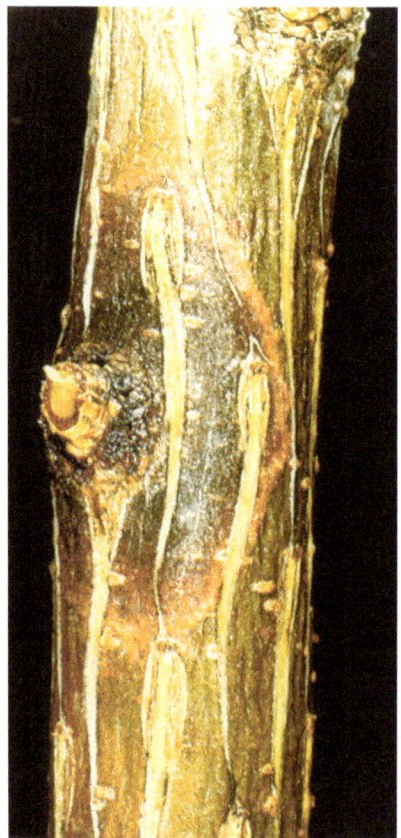

Figure 13.1—*Closeup of Phomopsis canker on a larch seedling.* Photo from USDA Forest Service Region Six Archives.

sunken because of callus tissue growth around the margins. Often a small dead branchlet is noticeable in the center of the canker. Cutting away bark at the edge of the canker reveals a sharp edge between healthy white and reddish or brown diseased tissue. Foliage beyond the canker turns yellow and dies quickly after the canker encircles the twigs or branches. When the canker occurs in succulent young growing tissue, rapid tissue death may cause the stem to bend into the shape of a shepherd's crook (fig. 13.2). These symptoms are typical of many diseases, so observation of fruiting bodies and spores is necessary to positively identify the pathogen. Both the asexual and sexual states of the fungus produce fruiting bodies on dead tissue within the cankers. They are visible with a 10x hand lens and appear as small, black, spherical pimples (fig. 13.3). *P. lokoyae* and *P. occulta* produce two types of single-celled spores; ellipsoid alpha-spores, and needlelike, curved, beta-spores. The alpha-spores of both species are similar in shape and size (6 to 10 by 2 to 4 microns), while the beta-spores of *P. lokoyae* are smaller (10 to 12 by 1.5 to 2 microns) than *P. occulta* (20 to 30 by 1 microns).

Figure 13.2—*Douglas-fir seedling with shepherd's crook caused by rapid death of succulent young tissue.* Photo by Katy M. Mallams, USDA Forest Service.

Biology

The asexual spores produced in the pycnidia are waterborne and spread short distances to new hosts by rain splash or sprinkler irrigation. Under favorable temperature and humidity conditions, the

Conifer Diseases

13. Phomopsis Canker

Figure 13.3—*Fruiting bodies of* Phomopsis *on the stem of a larch seedling.* Photo by Petr Kapitola, State Phytosanitary Administration, Czechia, at http://www.bugwood.org.

spores germinate and infect young shoots and small branchlets. Succulent shoots and tops of young seedlings are killed rapidly. In older seedlings with woody tissue, the fungus grows into and kills the inner bark of the branch or stem during the winter. Asexual fruiting bodies are produced the following spring and summer. Fruiting bodies of the sexual state are produced in autumn. These fruiting bodies produce windborne spores, which may play a role in long-distance spread. The fungus can also persist as a saprophyte on fallen cones and dead twigs, and in dead tissue on live seedlings.

Control

Cultural

Weakened trees are most susceptible to infection by canker fungi. In nurseries, extra attention to watering during unusually hot, dry periods may help prevent stress that predisposes seedlings to disease. Seedlings stressed by early fall or late spring frost, herbicide damage, and other abiotic events should also be carefully monitored. Wounds, including those caused by top-pruning, and fissures in the stems of Douglas-fir seedlings caused by rapid elongation, provide infection courts for this and other canker fungi. Preventing wounds and avoiding over-fertilization that results in rapid top growth may reduce the likelihood of infection.

Chemical

Phomopsis canker is normally so scattered that chemical use is not necessary. However, when conditions are especially favorable for the fungus, regular application of fungicides may prevent infection and disease development.

Selected References

Funk, A. 1973. Phomopsis (Diaporthe) canker of Douglas-fir in British Columbia. Forest Insect and Disease Survey Pest Leaflet No. 60. Pacific Forest Research Centre, Canadian Forestry Service, Victoria, British Columbia, Canada. 5 p.

Haase, D.; Khadduri, N.; Landis, T. 2007. Stem splitting and cankering in Pacific Northwest Douglas-fir seedlings. Tree Planters' Notes. 52(1): 9–11.

Hansen, E.M. 1990. Phomopsis canker. In: Hamm, P.B.; Campbell, S.J.; Hansen, E.M., eds. Growing healthy seedlings: identification and management of pests in Northwest forest nurseries. Special publication 19. Corvallis, OR: Oregon State University Forest Research Laboratory. 110 p.

Kliejunas, J.T.; Smith, Jr., R.S. 1989. Phomopsis canker of Douglas-fir. In: Cordell, C.E.; Anderson, R.A.; Hoffard, W.H.; Landis, T.D.; Smith, Jr., R.S.; Toko, H.V., tech. coords. Forest nursery pests. Agriculture Handbook 680. Washington, DC: USDA Forest Service: 59–61.

Thomson, A.J.; Trotter, D.D.; Shaykewich, D.; Banfield, R. 2009. Diseases and insects in British Columbia forest seedling nurseries. Phomopsis canker and foliage blight. http://forestry-dev.org/diseases/nursery/pests/phomopsi_e.html.

Conifer Diseases

14. Pitch Canker of Pines
Scott A. Enebak, Tom Starkey, and Tom Gordon

Revised from chapter by George M. Blakeslee, Thomas Miller, and Edward L. Barnard, 1989.

Hosts

Pitch canker, caused by the fungus *Fusarium circinatum* (syn. *Fusarium subglutinans F. moniliforme* var. *subglutinans*), has been reported on over 20 pine species throughout the world. In the Southern United States, hosts include slash, loblolly, longleaf, shortleaf, sand, spruce, and Virginia pine. In nursery settings, the disease is most serious on slash, longleaf, and shortleaf pine. In California, Monterey, knobcone, and bishop pines are the most widely affected host species. Monterey pines can suffer significant damage in seedling nurseries.

Distribution

This disease was first detected in the Eastern United States during the 1940s and can now be found throughout the Southern United States from Virginia to Florida, west to Texas in any natural forest, plantation, seed orchard, and forest tree nurseries that produce susceptible pine hosts. The fungus also has been introduced into California where the pathogen, with help from a suite of insect vectors, infects and kills susceptible pines in landscapes, native forests, Christmas tree farms, and seedling nurseries.

Damage

This disease can cause significant seedling mortality in the nursery and may appear as pre- and post-emergence damping-off or seedling culls at the end of the growing season. Also, seedlings that appear healthy but have latent infections may result in seedling mortality after outplanting and thus, additional losses may occur in the field.

Diagnosis

On young seedlings, lesions either on the stem at the groundline or on the upper taproot result in foliage discoloration and seedling death. Seedlings may either remain erect or collapse and be misdiagnosed as damping-off. Seed infestation by *F. circinatum* should be strongly suspected if poor germination and low stand densities are associated with individual seed sources. On older seedlings, stem lesions result in a purplish discoloration, followed by top dieback. Stem lesions near the groundline or the upper taproot usually cause discoloration and foliage wilt on the upper stem (fig. 14.1). If large enough, removal of the bark along the seedling stem or taproot of the seedling may reveal the darkly stained, resin-soaked xylem, which is a symptom of fungal infection (fig. 14.2).

The disease can be easily diagnosed later in the growing season as either single dead seedlings (seed infestation) or small groups (soil infestation) of seedlings found scattered throughout nursery beds (fig. 14.3). Look for discolored, yellow-green, brown to red foliage. On seedlings with succulent tissues, look for wilting foliage along the stem. Wilting results from the development of resin-soaked lesions either on the stem, near the groundline or on the upper portion of the taproot. Removing the bark exposes the resin-soaked wood, which is the primary method to confirm the presence of pitch canker when compared to a healthy, uninfected seedling (fig. 14.4). Particularly with longleaf pine, seemingly disease-free healthy seedlings with latent infections are lifted, only to die after outplanting in the field.

The fungus can be cultured from infected tissues on acidified potato dextrose

Figure 14.1—*Discoloration and foliage wilt of Monterey pine seedlings infected by* Fusarium circinatum. Photo by Cassandra Swett, University of California.

Figure 14.2—*Stained, resin-soaked xylem of a Monterey pine seedling infected by* Fusarium circinatum. Photo by Cassandra Swett, University of California.

Conifer Diseases

14. Pitch Canker of Pines

Figure 14.3—*Sand pine seedling affected by pitch canker.* Photo by George Blakeslee, University of Florida.

agar. The pathogen is characterized microscopically by curved, multiseptate macroconidia, 32 to 53 by 3.0 to 4.5 microns (fig. 14.5) and abundant oval to oblong microconidia, 8 to 12 by 2 to 3 microns, produced on polyphialides (fig. 14.6); chlamydospores are absent. The fungus also produces coiled knots of hyphae that do not produce spores.

Biology

The most common mode of entry into the nursery is via infested seed. The fungus can also enter a nursery as airborne inoculum from nearby infected trees or can possibly be carried by insects. The distribution of infected seedlings within nurseries suggests that insects may be an important wounding agent. After a nursery becomes infested, secondary disease spread during the growing season probably results from inoculum produced on previously infected seedlings. Insects and their feeding activities may play a role in late-season infections.

Control

Prevention

Use disease-free seed to prevent the fungus from being introduced into the nursery and causing seedling mortality. Suspected seedlots should be tested to determine if the seed is infested. Infested seedlots should not be sown or at a minimum, should be disinfested using fungicidal seed treatments. Suspected seed or seed of unknown origin should not be mixed with other noninfested seedlots during processing and seed treatment. Within pine species, families can differ in susceptibility and some may

Figure 14.4—*Resin-soaking at the root collar of longleaf pine caused by* Fusarium circinatum *(left) and a healthy uninfected seedling (right).* Photos by Tom Starkey, Auburn University.

Figure 14.5—*Macroconidia of* Fusarium circinatum. Photo by Michelle M. Cram, USDA Forest Service.

Conifer Diseases

14. Pitch Canker of Pines

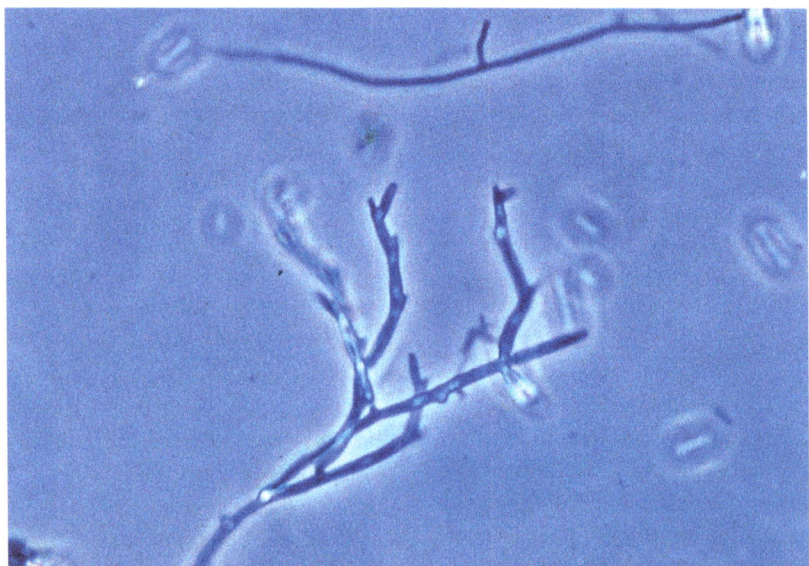

Figure 14.6—*Conidiophores (polyphialides) and microconidia of* Fusarium circinatum. Photo by Michelle M. Cram, USDA Forest Service.

sustain visible evidence of disease in seed orchards. Cones from those families should not be processed with disease-free cones. Proper cone and seed-handling procedures must be used to minimize fungal contamination of seed in the cone collection process.

Cultural

Remove pitch canker-infected trees in windbreaks, seed orchards, border plantings, or adjacent stands, thereby reducing nearby inoculum sources that could enter the nursery. Sanitation of seedling beds during the growing season may be achieved by removing and destroying infected seedlings from the nursery to prevent fungal spread. Care should be taken not to move soil associated with diseased trees into noninfested areas of the nursery. Soil treatment at infested sites with an eradicative material may be appropriate (see Chemical section that follows).

During lifting and packing, cull symptomatic seedlings to reduce dispersal of the fungus to outplantings and to minimize exposure of healthy seedlings to diseased ones. Culling is particularly important with longleaf pine that may be infected, but do not exhibit the symptoms in the nursery. Removing bark from seedling stems will expose the darkly stained, resin-soaked wood, which is a diagnostic symptom of *Fusarium circinatum* infection.

Chemical

Fumigate with standard formulations and dosage rates to eradicate the pathogen from the soil. Treating the seed with hydrogen peroxide or bleach may decrease the chances of seed infestations and subsequent seedling mortality by this fungus. The treatment of seed and timely applications of fungicides registered for the fungus will decrease the disease incidence. Registered efficacious fungicides for pitch canker control are limited. When an outbreak of the disease is observed, insect control (especially fungal gnats in containers) will help minimize the spread of the pathogen.

Selected References

Aegerter, B.J.; Gordon, T.R.; Storer, A.J.; Wood, D.L. 2003. Pitch canker: a technical review. University of California Agriculture and Natural Resources Publication 21616. 13 p.

Barnard, E.L.; Blakeslee, G.M. 1980. Pitch canker of slash pine seedlings: a new disease in forest tree nurseries. Plant Disease. 64: 695–696.

Blakeslee, G.M.; Dwinell, L.D.; Anderson, R.L. 1980. Pitch canker of southern pines: identification and management considerations. For. Rep. SA FR11. Atlanta, GA: USDA Forest Service, Southern Area, State and Private Forestry. 15 p.

Blakeslee, G.M.; Miller, T.; Barnard, E.L. 1989. Pitch canker of southern pines. In: Cordell, C.E.; Anderson, R.L.; Hoffard, W.H.; Landis, T.D.; Smith, Jr., R.S.; Toko, H.V., tech. coords. Forest nursery pests. Agriculture Handbook 680. Washington, DC: USDA Forest Service: 64–65.

Carey, W.A.; Kelley, W.D. 1994. First report of *Fusarium subglutinans* as a cause of late-season mortality in longleaf pine nurseries. Plant Disease. 78: 754.

Carey, W.A.; Oak, S.W.; Enebak, S.A. 2005. Pitch canker ratings of longleaf pine clones correlate with *Fusarium circinatum* infestation of seeds and seedling mortality in containers. Forest Pathology. 35: 205–212.

Starkey, T.S.; Enebak, S.A. 2010. The use of prothioconazole to control forest nursery diseases of *Pinus* spp. In: Proceedings of the 7th meeting of IUFRO Working Party 7.03-04. Report 10-01-01. USDA Forest Service. Southern Region, Forest, Health Protection: 92–103.

Conifer Diseases

15. Rhizoctonia Blight of Southern Pines
Tom Starkey and Scott A. Enebak

Revised from chapter by James T. English and Edward L. Barnard, 1989.

Hosts

Rhizoctonia blight, caused by species of *Rhizoctonia* (sexual states in the genera *Thanatephorus* or *Ceratobasidium*) occurs on many southern pine species. In forest seedling nurseries, longleaf pine is highly susceptible to Rhizoctonia blight due to the close proximity of the needles to the soil, while loblolly pine is more affected by aerial web blight. Slash pine has been observed to be resistant to the aerial form of the disease.

Distribution

Rhizoctonia blight occurs throughout the Southern United States wherever susceptible hosts are grown.

Damage

Rhizoctonia blight can cause significant seedling mortality in both bareroot and container nurseries growing longleaf and loblolly pine. Seedling damage by *Rhizoctonia* includes damping-off and rot of roots, stems, needles, and terminal buds. The early grass stage of longleaf pine is particularly vulnerable to damping-off, needle loss, rot, and eventual mortality. On loblolly pine, aerial web blight is the most common form of damage. The disease does not generally cause widespread mortality across nursery beds, but rather occurs in isolated disease foci within a nursery bed that may coalesce over time (fig. 15.1). Coastal loblolly pine families are more susceptible to the aerial web blight form although the disease severity varies among pine families. The effects of infection in the nursery (without mortality) on field performance of outplanted stock are unknown.

Diagnosis

On longleaf pine, water-soaked, chlorotic lesions appear at the base of needles (fig. 15.2). The basal ends of the needles initially appear healthy but gradually

Figure 15.1—*Rhizoctonia foliar blight disease foci in loblolly pine (foreground).* Photo by Tom Starkey, Auburn University.

turn yellow and then brown. In time, the needle base, terminal bud, and upper taproot, darken and decay (fig. 15.3). Foliar aerial web blight on loblolly pine can become severe among the lower, interior needles within a nursery bed weeks before symptoms even become visible on the seedling's upper foliage. Often, the disease is discovered only when upper needles are top-clipped, revealing brown patches, dead needles, and bare stems within the canopy (fig. 15.4). Infected foliage turns gray and is covered with fine aerial mycelial webbing. The affected foliage abscises, leaving a bare seedling stem. Unlike longleaf pine, loblolly pine stems and buds do not appear to be infected by the fungus. Frequently, Rhizoctonia blight appears in circular to irregular patterns scattered within the seedling bed. These infection centers typically consist of several dead seedlings surrounded by seedlings with different degrees of discoloration and stages of infection (fig. 15.5). *Rhizoctonia* species

Figure 15.2—*Necrotic and chlorotic lesion at the base of longleaf pine associated with* Rhizoctonia *species.* Photo by Edward L. Barnard, Florida Division of Forestry.

Conifer Diseases

15. Rhizoctonia Blight of Southern Pines

Figure 15.3—*Death of terminal bud (left) and needle bases of longleaf pine caused by* Rhizoctonia *species. Healthy seedling is shown on the right.* Photo by Edward L. Barnard, Florida Division of Forestry.

Figure 15.4—*Necrotic lower needles associated with Rhizoctonia foliar blight in loblolly pine.* Photo by Tom Starkey, Auburn University.

machinery, or mulch material added to the nursery soils. Longleaf pine seedlings are infected through the terminal bud and needle base at or just below the soil surface. Irrigation and rain water, which splash soil onto the low-growing longleaf pine seedlings, create conditions favorable to infection. This situation is often intensified in sandy soils. After a nursery bed becomes infested, the fungus spreads within soil by mycelial growth. The exact mode of infection for the aerial blight in loblolly pine is unknown, but infection is initiated on the weakened foliage within the interior seedling canopy during warm weather and free moisture after seedling canopy closure within a nursery bed. Basidiospores may also be involved with infection as well. *Rhizoctonia* overwinters as sclerotia, either within plant debris or in the soil. The disease tends to be more severe in soils that are in their second and third crop since soil fumigation.

can be cultured easily on most common laboratory media. Cultures vary in color from pale yellow to dark brown. These fungi generally do not produce spores in culture. However, cultures develop gray to dark brown sclerotia (fig. 15.6). Microscopically, these fungi may be recognized by the characteristic right-angle branching of the hyphae, which have constrictions at the points where they connect with parent hyphae (fig. 15.7). In older cultures, cross walls usually develop just beyond the hyphal constrictions.

Biology

How *Rhizoctonia* species infest a nursery is uncertain, but inoculum sources may include contaminated seed, airborne basidiospores, infested soil carried on

Figure 15.5—*Typical nursery bed damping-off of longleaf pine caused by* Rhizoctonia *species.* Photo by Edward L. Barnard, Florida Division of Forestry.

Conifer Diseases

15. Rhizoctonia Blight of Southern Pines

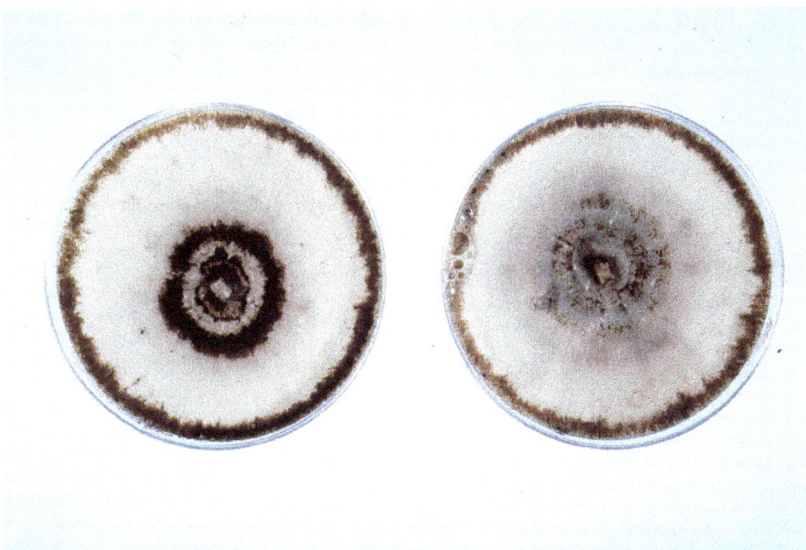

Figure 15.6—*Typical culture of* Rhizoctonia *spp. showing characteristic brown sclerotia.* Photo by Edward L. Barnard, Florida Division of Forestry.

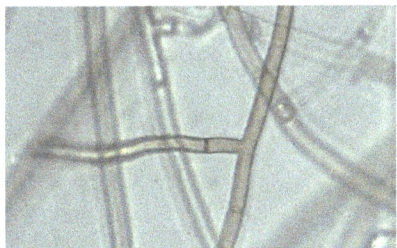

Figure 15.7—*Hyphae of* Rhizoctonia *species showing characteristic right-angle branching and constrictions near septa.* Photo by Gerald Holmes, at http://www.bugwood.org.

Control

Prevention

Use only recently collected, clean seed for sowing. When feasible, sow longleaf pine seed in the fall—losses in fall-sown longleaf pine have been appreciably less than when seedlings are sown in the spring. Variation exists among loblolly pine families and susceptibility to infection, piedmont sources tend to tolerate the fungus over coastal sources. In addition, Rhizoctonia blight appears to develop more rapidly in neutral to alkaline soils, so lower soil pH may mitigate the disease.

Cultural

Avoid the movement of infested soils within nurseries on machinery or hand tools. Irrigate early in the day to prevent long periods of wet foliage. Maintain soil pH below 6.0 to minimize activity of the fungus. High nitrogen fertilization favors *Rhizoctonia* development. Longleaf seed beds can be mulched to reduce soil splash.

Chemical

Fumigate nursery beds to reduce soilborne inoculum. Early detection of damping-off is critical to minimize seedling loss. After symptoms are observed, applications of a registered fungicide should begin as soon as possible. For the aerial blight, regular fungicidal sprays after canopy closure occurs can provide effective control to this disease. Efficacy of chemical control varies with chemicals.

Selected References

Davis, W.C. 1941. Damping-off of longleaf pine. Phytopathology. 31: 1011–1016.

English, J.T.; Barnard, E.L. 1989. In: Cordell, C.E.; Anderson, R.L.; Hoffard, W.H.; Landis, T.D.; Smith, Jr., R.S.; Toko, H.V., tech. coords. Forest nursery pests. Agriculture Handbook 680. Washington, DC: USDA Forest Service: 66–67.

English, J.T.; Ploetz, R.C.; Barnard, E.L. 1986. Seedling blight of longleaf pine caused by a binucleate Rhizoctonia solani-like fungus. Plant Disease. 70: 148–150.

Gilly, S.P.; Barnard, E.L.; Schroeder, R.A. 1985. Field trials for the control of Rhizoctonia blight of longleaf pine seedlings: effects of seedbed planting densities, fungicides, and mulches. In: South, D.B., ed. Proceedings of the international symposium on nursery management practices for the southern pines; 1985 August 4-5; Montgomery, AL. Montgomery, AL: Alabama Agriculture Experiment Station, Auburn University, and the International Union of Forest Research Organizations: 476–485.

Parmeter, Jr., J.R., ed. 1970. *Rhizoctonia solani*: biology and pathology. Berkeley: University of California Press. 255 p.

Papavizas, C.G.; Davey, C.B. 1961. Saprophytic behavior of *Rhizoctonia* in soil. Phytopathology. 51: 693–699.

Runion, G.B.; Kelley, W.D. 1993. Characterization of a binucleate *Rhizoctonia* species causing foliar blight of loblolly pine. Plant Disease. 77: 754–755.

Starkey, T.S.; Enebak, S.A. 2010. The use of prothioconazole to control forest nursery diseases of *Pinus* spp. In: Procceedings of the 7th meeting of IUFRO Working Party 7.03-04. Report 10-01-01. USDA Forest Service. Southern Region, Forest Health Protection: 92–103.

Conifer Diseases

16. Scleroderris Canker
Michael E. Ostry and Jennifer Juzwik

Revised from chapter by Darroll D. Skilling, 1989.

Hosts

Scleroderris canker, caused by the fungus *Gremmeniella abietina* (anamorph *Brunchorstia pinea*) consisting of several races and ecotypes, affects many conifer species. Two races of the fungus are known in North America. The North American race primarily affects jack and red pine in the Eastern United States and Canada and lodgepole pine in Western Canada. Austrian and eastern white pine are less commonly affected in the eastern region. The European race infects all pine species, but is primarily found on red and Scots pine and occasionally on Austrian and eastern white pine. The European race of the fungus is more aggressive than the North American race on these common hosts.

Distribution

The European race has been found in New York, Vermont, New Hampshire, Maine, southern Ontario, Quebec, New Brunswick, and Newfoundland. The North American race has been found in the Lake States, New York, and New England. In Canada, the North American race is widely distributed in Ontario, Quebec, Nova Scotia, and New Brunswick, and has occasionally been found in British Columbia and Alberta on lodgepole pine.

Damage

The disease can cause significant mortality to susceptible hosts in the nursery. The risk of spreading the fungus through nursery stock shipments is high because symptoms are frequently not visible until the spring after lifting and shipping even though infection occurred during the previous season. Infected seedlings die after outplanting, but also serve as foci for further spread of the fungus. Outplanting of pine seedlings infected with either race may result in large losses in the field. Detection of diseased seedlings within the nursery may result in curtailment of seedling shipments and potential destruction of all similar nursery stock in the same field.

Diagnosis

Both races of the fungus cause similar symptoms on infected seedlings and the races cannot be distinguished by the morphological spore characteristics. Look for an orange-brown discoloration at the infected needle base (fig. 16.1). Infected needles later turn brown. Often infected needles are loose and drop during late spring and summer leaving only the bare stem. Infected buds die and, therefore, do not expand in the spring. Resinous lesions and a yellow-green discoloration under the bark are sometimes observed in recently killed tissue. Small, black fruiting structures called pycnidia (1 mm wide) are commonly produced at the base of dead needles on dead stems (fig. 16.2). The spores (conidia) produced in pycnidia have 4 to 5 cells, pointed ends and are 30 by 3 microns in size. Microconidia (5.0 by 1.5 microns) are occasionally produced in some pycnidia.

Biology

The fungus is spread by airborne or rain-splashed spores. The North American race produces both asexual (conidia) and sexual spores (ascospores), while ascospores of the European race are rare. Ascospores are produced by dark brown apothecia; 1 mm wide, cup-like structures that form on dead infected twigs. Ascospores can initiate the disease in the nursery, but the asexual stage of the pathogen causes the greatest spread and damage.

Infection can take place throughout the growing season, but the primary infection period is from May to July. The fungus can grow at low temperatures in

Figure 16.1—*Orange discoloration at base of needles infected by* Gremmeniella abietina. Photo from USDA Forest Service Archive.

Conifer Diseases

16. Scleroderris Canker

Figure 16.2—*Pycnidia of* Gremmeniella abietina *at bases of infected needles on killed stems.* Photo from USDA Forest Service Archive.

the winter down to -6 °C (21.2 °F) and becomes especially damaging under heavy snow when it can extensively colonize seedlings.

Control

Sclerroderris canker is easily controlled in the nursery with regular fungicide applications. Begin treatment as soon as new growth appears in the spring. Sprays should be repeated at 2-week intervals until the first of July, then at 4-week intervals until September. One or two extra sprays may be needed if rainfall is unusually heavy during early summer. The European strain may require spraying every 2 weeks until late October because of a longer spore dispersal period.

Selected References

Manion, P.D., ed. 1984. Scleroderris canker of conifers. The Hague, The Netherlands: Martinus Nijhoff/W. Junk Publishers: 273 p.

Marosy, M.; Patton, R.F.; Upper, C.D. 1989. A conducive day concept to explain the effect of low temperature on the development of Scleroderris shoot blight. Phytopathology. 79: 1293–1301.

Petrini, O.; Petrini, L.E.; Laflamme, G; Ouellette, G.B. 1989. Taxonomic position of *Gremmeniella abietina* and related species: a reappraisal. Canadian Journal of Botany 67: 2805–2814.

Sinclair, W.A.; Lyon, H.H. 2005. Diseases of trees and shrubs, 2nd ed. Ithaca, NY: Cornell University Press. 660 p.

Skilling, D.D. 1989. Scleroderris Canker. In: Cordell, C.E.; Anderson, R.L.; Hoffard, W.H.; Landis, T.D.; Smith, Jr., R.S.; Toko, H.V.; tech. coords. Forest nursery pests. Agriculture Handbook 680. Washington, DC: USDA Forest Service: 69–70.

Skilling, D.D.; Schneider, B.; Fasking, D. 1986. Biology and control of Scleroderris canker in North America. Res. Pap. NC-275. USDA Forest Service. 18 p.

Conifer Diseases

17. Sirococcus Shoot Blight
Glen R. Stanosz

Hosts

Sirococcus shoot blight is attributed to three fungi previously referred to as the single species *Sirococcus conigenus* (syn. *S. strobilinus*). The species now known as *S. conigenus* most commonly damages pines, including red pine and lodgepole pine, and spruces such as Colorado blue spruce and Sitka spruce, but other conifers can also be diseased. Hemlocks and true cedars are reported *S. tsugae* hosts. *Sirococcus piceae* is known only from limited collections of spruce, and pathogenicity is yet to be confirmed.

Distribution

S. conigenus is reported from various locations in the Eastern, North-Central, and Northwestern regions of the United States, but its distribution within this range is likely discontinuous. *S. tsugae* is confirmed from the Pacific Northwest and both the Northeastern and Southeastern United States. To date, *S. piceae* has been found in both Eastern and Western Canada, but occurrence in the United States is not known.

Damage

Both *S. conigenus* and *S. tsugae* can damage seedlings of all ages, saplings, and larger ornamental and forest trees. Blighting may result in death or otherwise render seedlings unmerchantable. *S. conigenus* can be seedborne and also has been detected from asymptomatic seedlings.

Diagnosis

Infection of current year's growth can result in rapid shoot mortality, with either needle loss or dead needle retention, dependent on host and pathogen species (figs. 17.1 and 17.2). On pines, purplish stem lesions can expand to girdle and kill shoots, with curling or crooking of stems. On red pine saplings and larger trees, needles often droop near their base, turn brown to gray, and remain attached to dead shoots into the next year (fig. 17.3). Symptoms often are limited to the current year's shoot growth, or current and previous year's growth, without progression into older organs.

Asexual *Sirococcus* fruiting bodies are black flask-shaped pycnidia that can be seen with the naked eye or a hand lens. They are produced in dead needles and stems, and also on open female cones. They may be solitary or in groups, and erupt through the epidermis and cuticle (fig. 17.4). Pycnidia are sometimes numerous on needle bases and below the fascicle sheaths of dead pine needles, but on spruces may be more abundant on dead stems. Although *Sirococcus* species conidia are morphologically similar, spore examination does allow

Figure 17.1—*Needle loss and death of shoot tips of mountain hemlock infected by* Sirococcus tsugae. Photo by Glen R. Stanosz, University of Wisconsin-Madison.

Figure 17.2—*Jack pine seedlings with shoots killed by* Sirococcus conigenus. Photo by USDA Forest Service, North Central Research Station Archive, at http://www.bugwood.org.

Conifer Diseases

17. Sirococcus Shoot Blight

Figure 17.3—*Sirococcus shoot blight needle droop symptoms on red pine.* Photo by Glen R. Stanosz, University of Wisconsin-Madison.

differentiation of Sirococcus shoot blight from other diseases with similar symptoms, such as Diplodia shoot blight. Conidia of these three *Sirococcus* species are colorless, fusiform, with rounded tips and rounded to slightly truncate bases, two-celled, and approximately 9 to 16 by 2 to 4 microns in size (fig. 17.5).

Sirococcus species cultures can be obtained by placing pieces of surface-disinfested, symptomatic needles or stems on malt extract agar or potato dextrose agar amended with lactic acid or streptomycin sulfate to inhibit bacterial growth. Pycnidium and conidium production may be stimulated by placing sterile host needles on the medium, with incubation in the light at 20 °C to 24 °C (68 °F to 75 °F). Molecular methods can be used to identify isolates of each of the three *Sirococcus* species associated with conifers, or detect their presence on or in host samples without the need to obtain cultures.

Biology

Sirococcus conifer pathogens survive in and sporulate on dead needles, stems, cones on diseased trees, and debris on the ground. Viable spores can be disseminated by rain splash year-round, but are most abundant during spring and early summer when young shoots are most susceptible. Spores can be splashed from seedling to seedling in nursery beds, and from overstory trees to understory seedlings and saplings. Moist weather and low light conditions reportedly favor infection.

Control

Biological

Grow nonhost species in areas of nurseries where inoculum is available.

Cultural

Eliminate inoculum sources, including host trees in windbreaks and adjacent forests, in the nursery vicinity to minimize disease. Host materials, such as twigs, needles, and cones, should not be used as soil amendments or mulches. Practices that promote shoot drying, such as early morning irrigation and decreasing bed densities, may reduce infection frequency. Do not move infested seed or diseased seedlings into or out of the nursery.

Chemical

Protectant fungicides can reduce disease incidence in nursery beds. Repeated sprays during shoot elongation may be required, however.

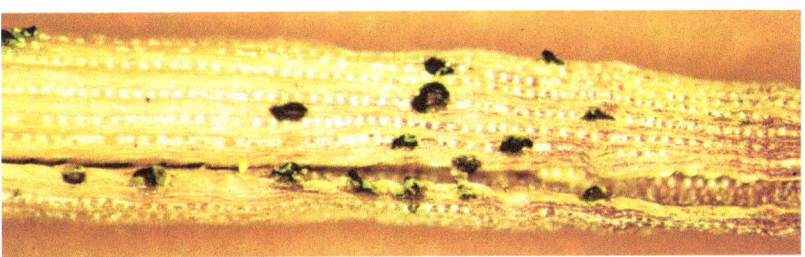

Figure 17.4—*Pycnidia of* Sirococcus conigenus *on pine needles.* Photo by USDA Forest Service, North Central Research Station Archive, at http://www.bugwood.org.

Conifer Diseases

17. Sirococcus Shoot Blight

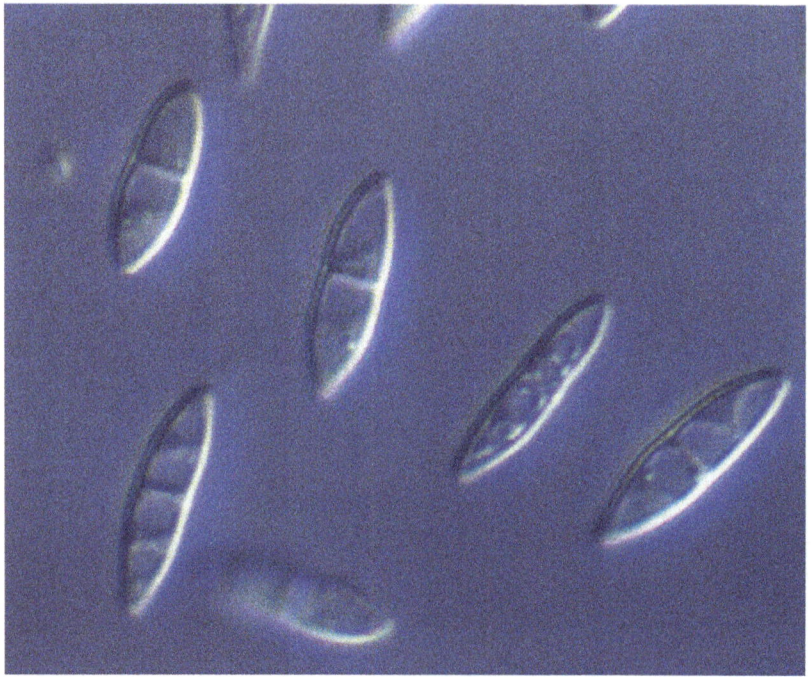

Figure 17.5—*Conidia of* Sirococcus conigenus. Photo by Glen R. Stanosz, University of Wisconsin-Madison.

Selected References

Miller-Weeks, M.; Ostrofsky, W. 2010. *Sirococcus tsugae* tip blight on eastern hemlocks. Pest alert. NA–PR–01–10, 2010. Newtown Square, PA: USDA Forest Service. 2 p.

Rossman, A.Y.; Castlebury, L.A.; Farr, D.F; Stanosz, G.R. 2008. *Sirococcus conigenus*, *S. piceicola*, sp. nov. and *S. tsugae* sp. nov. on conifers: anamorphic fungi in the *Gnomoniaceae*, *Diaporthales*. Forest Pathology. 38: 47–60.

Smith, D.R.; Stanosz, G.R. 2008. PCR primers for identification of *Sirococcus conigenus* and *S. tsugae*, and detection of *S. conigenus* from symptomatic and asymptomatic red pine shoots. Forest Pathology. 38: 156–168.

Smith, Jr., R.S.; Nicholls, T.H. 1989. Sirococcus shoot blight. In: Cordell, C.E.; Anderson, R.L.; Hoffard, W.H.; Landis, T.D.; Smith, Jr., R.S.; Toko, H.V., tech. coords. Forest nursery pests. Agriculture Handbook 680. Washington, DC: USDA Forest Service: 71–72.

Conifer Diseases

18. Snow Molds of Conifers
Jill D. Pokorny

Hosts

Snow molds are a select group of fungi that grow and attack dormant plants at low temperatures under snow cover. They are fungal pathogens of forage crops, winter cereals, and conifers. These fungi have adapted different survival strategies to enable them to grow at subzero temperatures under snow. The freezing resistance of mycelia and spores is considered a key to their survival.

Two types of snow molds affect conifers: snow blights and brown felt blights. Snow blight fungi produce annual mycelium that disappears from plant surfaces soon after snowmelt, and brown felt blight fungi produce brown mycelium that grows perennially on plant surfaces and persists long after snowmelt.

Two fungi account for most of the snow blight records in North America: *Phacidium abietis* and *Lophophacidium hyperboreum*. Other snow blight fungi, most notably, *Phacidium infestans* and *Sarcotrochila*, have been cited to occur in North America; however, these reports are considered unconfirmed due to changes in taxonomic nomenclature and questions regarding pathogenicity.

Two fungi commonly cause brown felt blight in North America: *Herpotrichia juniperi* and *Neopeckia coulteri*. *Herpotrichia juniperi* occurs on fir and spruce and *Neopeckia coulteri* occurs only on pines.

Snow blights affect many conifer hosts including Engelmann, Norway, black, Colorado blue, red, and white spruces; balsam, grand, white, Pacific silver, and subalpine firs; Douglas-fir; eastern and western white pine; and eastern hemlock.

Brown felt blights also affect many conifer hosts including Engelmann, Norway, Sitka, and white spruces; alpine, balsam, grand, noble, red, European silver, Pacific silver, and white firs; bristlecone, foxtail, Jeffrey, limber, lodgepole, mugo, ponderosa, sugar, western white, and whitebark pines; mountain and western hemlocks; western redcedar; common, creeping, alligator, and Rocky Mountain junipers; and Pacific yew.

Distribution

Snow blights occur in the boreal and alpine regions of the United States, especially in areas of higher elevations and with deep snow accumulation. Brown felt blights occur only in the mountains of the West.

Damage

Significant losses can occur to seedlings in northern nurseries when snow cover persists in the spring. Small trees, in natural and planted forests, can also be affected, resulting in severe damage to lower branches that are below the snow line. Small trees beyond the seedling stage will often retain some foliage and survive.

Diagnosis

Look for severe browning or mortality of seedlings in the early spring, just after snowmelt (fig. 18.1).

Because snow mold fungi can grow under snow, seedlings that appear healthy in the fall may show severe damage the next spring. In nursery beds, infection frequently occurs in patches and is usually most severe in areas where snowmelt is delayed. Needles, twigs, and branches become covered with a mass of mycelium (mycelial mats) that become visible as the snow melts. The mycelial mat color and persistence are diagnostic features that help to distinguish between brown felt blights and snow blights.

Brown felt blights are readily identifiable by the brown mycelial mats that cover twigs, branches, and needles (fig. 18.2). Fungal growth is prolific on

Figure 18.1—*White spruce seedlings in a nursery bed showing symptoms of snow blight.* Photo by Gaston Laflamme, Canadian Forest Service, Laurentian Forestry Centre, as published in *Diseases of Trees and Shrubs*, 2nd edition, Cornell University Press, 2005.

Figure 18.2—*Alpine fir showing the diagnostic dark mycelial mats of brown felt blight.* Photo by Cornell University Department of Plant Pathology and Plant-Microbe Biology, as published in *Diseases of Trees and Shrubs*, 1st edition, Cornell University Press, 1987.

Conifer Diseases

18. Snow Molds of Conifers

branches and foliage that are buried under snow (fig. 18.3). As the snow melts, and mats are exposed, fungal development ceases. Freshly exposed mats are dark brownish-black, weathering to grayish-brown. They are resistant to summer desiccation, persist on infected needles, and resume growth during the winter and the next spring as a perennial blight. The needles and twigs under the mats die but remain attached for 1 year or more. Spherical black fruiting bodies form on and in the mats the second winter after infection.

In the case of snow blights, affected needles become covered with mycelial mats that are white and cobweb-like in appearance (fig. 18.4). Unlike the brown mycelial mats of felt blights, the white mats of snow blights are ephemeral and do not persist after the snow melts. Affected needles first appear yellow, gradually turn red to reddish-brown, and finally to gray, and remain attached for 1 to 2 years. In late summer, black fruiting bodies appear on the underside of discolored needles. Morphology of these fruiting bodies varies with the fungus species involved. Fruiting bodies may appear as small black dots, aligned in rows on each side of the midrib (fig. 18.5) or as elongated black lines that extend along one-half or more of the length of the needle.

Biology

The biology of snow mold fungi is unique. Although most plant pathogenic fungi overwinter in a dormant state, the snow molds actively grow during the winter and infect conifer needles that are buried under the snow.

Black fruiting structures develop on blighted needles and release spores during moist weather in the fall. The spores are

Figure 18.4—*Snow blight mycelium and symptoms on white spruce needles recently under snow.* Photo by Gaston Laflamme, Canadian Forest Service, Laurentian Forestry Centre, as published in *Diseases of Trees and Shrubs*, 2nd edition, Cornell University Press, 2005.

Figure 18.5—*Fir needles showing snow blight fruiting bodies (small black dots), aligned in two lines on either side of the midrib.* Photo by Susan K. Hagle, USDA Forest Service, at http://www.bugwood.org.

windborne, land on healthy needles, and cause infection once the needles become covered in snow. As the snow melts in the spring, mycelium grows from the infected needles to healthy nearby needles, forming the characteristic mycelial mats. Fungal growth ceases during the summer and resumes again in the fall. The windborne spores are the most important source of new infections.

Control

Prevention

Snow molds may occur on the low branches of trees up to pole size and on spruces and firs that are larger than pole

Figure 18.3—*Prolific mat development on lower branches of white fir that were buried in snow.* Photo by Duane Mallams.

Conifer Diseases

18. Snow Molds of Conifers

size; therefore, inoculum sources within the nursery should be removed. Use northern seed sources, which are usually more resistant to these fungi than seed from southern areas. If possible, avoid nursery sites where snowmelt is delayed.

Cultural

Avoid growing susceptible tree species in nursery beds where snow accumulates in drifts. Remove infected seedlings and those adjacent, apparently healthy seedlings to reduce the spread of the disease from localized infections. Increase potassium fertilizer application when potassium is deficient because low potassium increases snow blight severity.

Chemical

No fungicides are specifically registered for snow blight or brown felt blight control.

Selected References

Anonymous. 2003. Forest insect and disease identification and management. USDA Forest Service, Northern Region; Idaho Department of Lands, Insect and Disease Control; Montana Department of State Lands, Division of Forestry. 223 p.

Boyce, J.S. 1961. Forest pathology, 3rd ed. New York: McGraw-Hill. 572 p.

Funk, A. 1985. Foliar fungi of western trees. Victoria, British Columbia: Natural Resources Canada, Canadian Forestry Service, Pacific Forestry Centre. 159 p.

Hepting, G.E. 1971. Diseases of forest and shade trees of the United States. Agriculture Handbook 386. Washington, DC: USDA Forest Service. 658 p.

Hoshino, T.; Xiao, N.; Tkachenko, O.B. 2009. Cold adaptation in phytopathogenic fungi causing snow molds. Mycoscience. 50: 26–38.

Naoyuki, M. 2009. Snow molds: a group of fungi that prevail under snow. Microbes Environ. 24: 14–20.

Reid, J.; Cain, R.F. 1962. Studies on the organisms associated with "snow blight" of conifers in North America. II. Some species of the genera *Phacidium*, *Lophophacidium*, *Sarcotrochila*, and *Hemiphacidium*. Mycologia. 54: 481–497.

Sinclair, W.A.; Lyon, H.H. 2005. Diseases of trees and shrubs, 2nd ed. Ithaca, NY: Cornell University Press. 660 p.

Skilling, D.D. 1989. Snow blight of conifers. In: Cordell, C.E.; Anderson, R.L.; Hoffard, W.H.; Landis, T.D.; Smith, Jr., R.S.; Toko, H.V., tech. coords. Forest nursery pests. Agriculture Handbook 680. Washington, DC: USDA Forest Service: 73–74.

Stone, J. 1997. Felt blights and snow blights. In: Hansen, E.M.; Lewis, K.J., eds. Compendium of conifer diseases. St. Paul, MN: American Phytopathological Society: 63–64.

Conifer Diseases

19. White Pine Blister Rust

Lee E. Riley and Judith F. Danielson

Hosts

White pine blister rust, caused by the fungal pathogen *Cronartium ribicola*, is an exotic, invasive disease that is native to Asia. It was introduced to both the east and west coasts of the United States in the early 1900s on infected eastern white pine nursery stock imported from Europe.

Blister rust has a complex life cycle that requires alternate hosts for spread of the disease. The life cycle includes five spore forms that most commonly alternate between five-needle pines and currant or gooseberry (*Ribes* spp.) leaves, although Indian paintbrush and snapdragon have recently been determined to serve as alternate hosts as well.

All five-needle pines are susceptible to the fungus, although some species are more susceptible than others. Of the species native to North America, western white, sugar, eastern white, and whitebark pines are most susceptible; limber and southwestern white pines are moderately susceptible; and the susceptibility of bristlecone pine is under investigation. Blister rust is the most important exotic pathogen of forest trees worldwide, and has been particularly devastating to large forested areas and forest stand structure in North America.

Distribution

After introduction from Europe on both coasts of the United States, the pathogen spread throughout the entire range of white pines in North America. Although less of a problem in the East, the disease is found from the Atlantic Provinces of Canada south through Georgia and as far west as Minnesota and Iowa. In Western North America, the pathogen is infecting white pine species from British Columbia and Alberta through the Intermountain West and Pacific Northwest, south through the Rocky Mountain Region, California, and the Southwest, and into Mexico, and is actively continuing its spread in elevation and latitude. It is most severe where conditions are cool and moist for extended periods in late summer and early fall.

Damage

Although young seedlings are more susceptible to infection by *C. ribicola* than older trees, mortality is rare in 1- or 2-year-old nursery stock because of the length of the disease cycle. If spores are present and conditions are optimal, seedlings may become infected during the first growing season. Some mortality may occur in the second year or in any hold-over or transplanted stock. Depending on the genotype, mortality is quite likely to occur in infected seedlings following outplanting. Blister rust is a major cause of mortality in regenerating and outplanted five-needle pines, making reestablishment of these species extremely difficult (fig. 19.1).

Figure 19.1—*Recent blister rust mortality on 13-year-old outplanted sugar pine in southwestern Oregon.* Photo by Judith F. Danielson, USDA Forest Service.

Conifer Diseases

19. White Pine Blister Rust

Diagnosis

Pines. The infection court is needles; spores send germ tube hyphae through stomata, forming mycelia that grow down into the stem cambium. Within a few months of infection, diagnostic yellow, orange, or red flecks may (or will) appear on the needles, sometimes with necrotic spots in the center (fig. 19.2). After initial infection, the fungal hyphae grow through the needle and into the branch, eventually into the stem of the seedling. The established mycelium forms an orange-brown canker on stem tissue that continues to develop and expand. A slightly swollen or cankered area may form on the seedling branch about 12 to 18 months after the initial infection. This area may be characterized by browning bark with yellowish discoloration at the border of the area, followed by a distinctive spindle-shaped swelling (fig. 19.3). Approximately 1 year following formation of the swelling, pycnial lesions that are characterized by orange to yellow blisters will form (fig. 19.4). These pycnia will eventually produce aeciospores that will infect the alternate host species.

Figure 19.2—*(A) and (B) Needle spots on western white pine seedlings infected with blister rust; (C) infected versus noninfected seedlings.* Photos A and B by Richard Sniezko, USDA Forest Service, photo C by Judith F. Danielson, USDA Forest Service.

Figure 19.3—*Stem cankers on (A) western white pine, (B) southwestern white pine, and (C) limber pine infected with blister rust.* Photos by Judith F. Danielson, USDA Forest Service.

Forest Nursery Pests 75

Conifer Diseases

19. White Pine Blister Rust

Figure 19.4—(A) Pycnial colonies on western white pine; (B) active stem canker on western white pine. Photo A by Richard Sniezko, USDA Forest Service, photo B by Robert Danchok, USDA Forest Service.

on plants grown as a nursery crop, will create a high-hazard or high-risk situation for five-needle pine seedlings growing in the nursery.

Biology

The spread of *C. ribicola* is favored by moist, cool conditions. It is much less of a problem in nurseries located in dry environments. In late summer, basidiospores are released from the infected host plants (most commonly *Ribes* spp.). They can infect pine needles of any age. Spore germination and infection require 48 hours of 100 percent relative humidity at temperatures below 20 °C (68 °F). During this period, spore germ tubes penetrate the needles through the open stomata.

Diagnosis of rust infection in the nursery may not be possible. Seedlings should be carefully monitored during lifting and packing, but infection may not be evident until seedlings have been outplanted.

***Ribes* species.** Although resistant varieties of *Ribes* species do exist, most are susceptible to rust infection. In spring, urediniospores can be found on the underside of infected *Ribes* leaves (fig. 19.5). These spores are characterized by yellow to orange blisters, which are seen easily without magnification. These urediniospores will reinfect the *Ribes* leaves throughout the summer, thereby creating a buildup of inoculum. In late summer to early fall, hair-like structures, called telial columns, will form from the old uredinial blisters (fig. 19.6). These structures eventually release basidiospores, the spore form that infects pines under the proper conditions. Infected *Ribes* plants, either in areas surrounding the nursery or

Figure 19.5—Ribes nigrum *infected with* Cronartium ribicola *in August.* Photo by Judith F. Danielson, USDA Forest Service.

Conifer Diseases

19. White Pine Blister Rust

Figure 19.6—*(A) and (B) Telial columns on* Ribes *species leaves.* Photos by Robert Danchok, USDA Forest Service.

During the spring following infection, fungal hyphae grow through the needles and into the branch to form the branch swelling or canker. About 1 year later, pycnia and pycniospores are formed on the canker, which lead to the development of aecia and aeciospores in the same tissue during the following year. Aeciospores are hardy, highly infective, and can be disseminated up to 500 km (300 mi) by wind to infect the alternate host. After the alternate host is infected, urediniospores are produced on the leaves, causing reinfection of the host and a buildup of inoculum.

In late summer to early fall, telial columns are formed on the alternate host leaves, in which the teliospores germinate and produce the basidiospores that infect the pines.

Rust infection can be transferred between *Ribes* during the uredinial stage of the fungus, but rust does not transfer between pines at any stage in the life cycle. The rust is an annual infection on its herbaceous hosts, and is perennial on its pine hosts.

Control

Prevention

Since basidiospores are relatively short-lived and usually infect pines within 100 m (328 ft) of origin, host plant removal in a 300-to-500 m (984-to-1,640 ft) area is recommended to reduce the rust hazard. *Ribes* eradication has been attempted throughout both the Eastern and Western United States where white pines are native. The program resulted in various success rates because one infected *Ribes* plant per hectare can provide the necessary inoculum for a serious disease problem. *Ribes* eradication programs were somewhat successful in the East, but much less so in the West.

The best alternative for disease prevention in the nursery is to sow seeds from resistant genotypes that have been tested in long-term rust resistance breeding programs. Seed orchards of eastern white pine, sugar pine, and western white pine have been developed and are continually being improved by government agencies in Canada and the United States from the survivors of breeding and artificial disease inoculation programs. Seeds from these orchards are producing reforestation stock with higher viability than wild woods run seeds. Because the incidence of resistance is very low in native pine populations, even progeny of these programs may show incomplete or low resistance to blister rust. Resistant genotypes, or genotypes displaying partial resistance, are available for most breeding zones in eastern and western white pines and sugar pines, with research currently underway to produce resistant whitebark, southwestern white, limber, and bristlecone pines. Because of the ability of *C. ribicola* to mutate, it is important to maintain a broad array of resistance mechanisms in outplanting stock, including slow-rusting mechanisms, for balance. Planting of infected seedlings, a majority of nonresistant seedlings, or seedlings of unknown provenance almost completely ensures the eventual failure of the planting.

Cultural

If diagnosis is possible, all infected seedlings should be removed before or during lifting and packing. If high value, older seedlings are infected and infection has not reached the stem of the seedlings, pruning to remove the infection can be moderately effective.

If *Ribes* species are also grown as a crop in the nursery, all plants should be closely monitored and infected seedlings or infected leaves removed as soon as possible.

Conifer Diseases

19. White Pine Blister Rust

Chemical

Fungicides in the ethylene-bisdithiocarbamate class of compounds have been found to be effective in controlling rust fungi. Triadimefon, a systemic chemical in the triazole class, has shown some longer term effectiveness in protecting seedlings from white pine blister rust infection.

Selected References

Agrios, G. 1997. Plant pathology, 4th ed. San Diego, CA: Academic Press. 635 p.

Berube, J.A. 1996. Use of triadimefon to control white pine blister rust. Forestry Chronicle. 72: 637–638.

Boyce, J.S. 1961. Forest pathology. New York: McGraw-Hill. 572 p.

Danchok, R.; Sharpe, J.; Bates, K.; Fitzgerald, K.; Kegley, A.; Long, S.; Sniezko, R.; Danielson, J. 2003. Operational manual for white pine blister rust inoculation at Dorena Genetic Resource Center. [Unpublished document]. Located at: Cottage Grove, OR: USDA Forest Service, Dorena Genetic Resource Center.

Goheen, D.; Goheen, E. 2000. Forest insects and diseases: Natural Resources Institute. [Unpublished document]. Located at: Central Point, OR: USDA Forest Service, Forest Health Protection, Pacific Northwest Region, and the Dorena Genetic Resource Center.

Hunt, R. 1997. White pine blister rust. In: Hansen, E.M.; Lewis, K.J., eds. Compendium of conifer diseases. St. Paul, MN: American Phytopathological Society: 26–27.

McDonald, G.I.; Hoff, R.J. 2001. Blister rust: an introduced plague. In: Tomback, D.F.; Arno, S.F.; Keane, R.E., eds. Whitebark pine communities: ecology and restoration. 193–220.

Patton, R.F.; Cordell, C.E. 1989. White pine blister rust. In: Cordell, C.E.; Anderson, R.L.; Hoffard, W.H.; Landis, T.D.; Smith, Jr., R.S.; Toko, H.V., tech. coords. Forest nursery pests. Agriculture Handbook 680. Washington, DC: USDA Forest Service: 75–77.

Conifer Insects

Conifer Insects

20. Cranberry Girdler
Art Antonelli

Revised from chapter by David L. Overhulser and Paul D. Morgan, 1989.

Hosts

The cranberry girdler (*Crysoteuchia topiaria*) in the family Crambidae belongs to a large group of turfgrass pests called sod webworms. As well as damaging turf, this insect commonly damages 2-0 nursery stock of Douglas-fir, noble fir, larch, and spruce. Other stock occasionally damaged includes 1-0, 3-0, and 2-1 Douglas-fir and 1-0 larch.

Distribution

The cranberry girdler has been observed in bareroot conifer nurseries in the Western United States.

Damage

Cranberry girdler larvae feed on the roots and lower stems of seedlings and in some cases may completely girdle and kill the plant. Damage is most likely to occur in nurseries adjacent to grass fields, which are prime habitats for this insect.

Diagnosis

On the lower seedling stem and taproot, look for patches where the bark and cortex have been removed (fig. 20.1). Damage to seedlings is usually noticed during lifting operations or when severely damaged seedlings change color in the fall. At that point, control is ineffective.

Figure 20.1—*Feeding damage of the cranberry girdler on lower stem and roots of Douglas-fir seedlings.* Photo by Thomas D. Landis, USDA Forest Service.

Biology

Adult moths (fig. 20.2) emerge in grass fields from May to July. Moths are visible during the day. They fly with quick, jerky movements for short distances. Female moths deposit eggs on and around nursery stock. Eggs hatch in 3 to 5 days, and larvae (fig. 20.3) feed in nursery beds from June to October, when they spin the cocoons and overwinter. Feeding by the late instars in August to October is what damages seedlings. Moth populations vary from year to year because of the effects of predation and disease on larval survival. Birds such as starlings, killdeer, sandpipers, and blackbirds feed on overwintering larvae. A naturally occurring soil fungus, *Beauveria bassiana*, also kills overwintering larvae.

Figure 20.2—*Male adult of cranberry girdler.* Photo by Ken Gray. Image courtesy of Oregon State University.

Figure 20.3—*Larva of cranberry girdler.* Photo by Ken Gray. Image courtesy of Oregon State University.

Control

Cultural

Avoid using cover crops that might provide host material for the cranberry girdler or other sod webworms. Cultivate or apply herbicides to noncrop areas to control weeds and grasses.

Chemical

Traps baited with an attractant and placed in grassy areas adjacent to the nursery can be used to monitor populations and to time insecticide applications for moth control.

Selected References

Cramshaw, W. 2004. Garden insects of North America. Princeton, NJ: Princeton University Press. 656 p.

Hollingsworth, C.S.; Antonelli, A.; Hirnyck, R., eds. 2009. Pacific Northwest insect management handbook. Corvallis, OR: Oregon State University Press. 698 p.

Kamm, J.A.; Morgan, P.D.; Overhulser, D.L.; McDonough, L.M.; Triebwasser, M.; Kline, L.N. 1983. Management practices for cranberry girdler (Lepidoptera: Pyralidae) in Douglas-fir nursery stock. Journal of Economic Entomology. 76: 923–926.

Kamm, J.A.; Robinson, R.R. 1974. Life history and control of sod webworms in grass seed production. Ext. Circ. 851. Corvallis, OR: Oregon State University Extension Service. 2 p.

Overhulser, D.L.; Morgan, P.D. 1989. Cranberry girdler. In: Cordell, C.E.; Anderson, R.L.; Hoffard, W.H.; Landis, T.D.; Smith, Jr., R.S.; Toko, H.V., tech. coords. Forest nursery pests. Agriculture Handbook 680. Washington, DC: USDA Forest Service: 86–87.

Tunnock, S. 1985. Suppression of cranberry girdler damage in beds of Douglas-fir seedlings, Coeur d'Alene Nursery, Idaho Panhandle National Forest. Rep. 85-4. Missoula, MT: USDA Forest Service, Northern Region. 7 p.

Conifer Insects

21. Pine Tip Moths
John T. Nowak

Hosts

Tip moths, *Rhyacionia* species (family: Tortricidae), are very common young pine seedling pests throughout the United States, particularly in the Southeastern United States. The Nantucket pine tip moth (*R. frustrana*) (fig. 21.1) is a ubiquitous pest of loblolly, shortleaf, Virginia, pitch, Scots and red pine in the Eastern United States, and ponderosa and Monterey pine in the Western United States, where it was accidentally introduced. Longleaf, slash, and eastern white pine are also occasionally infested, but are much less susceptible. The subtropical pine tip moth (*R. subtropica*) and pitch pine tip moth (*R. rigidana*) are also found in the East, but typically infest slash, longleaf, and pitch pine, respectively. The western pine tip moth (*R. bushnelli*) prefers ponderosa pine, but also infests other pine species.

Figure 21.1—*Adult Nantucket pine tip moth.* Photo by Chris Asaro, Virginia Department of Forestry.

Distribution

The Nantucket pine tip moth's range extends from Massachusetts south to central Florida and west through central Missouri, Oklahoma, and east Texas. Disjunctive populations also exist in New Mexico, Arizona, and California. The Nantucket pine tip moth's eastern range overlaps with the Subtropical pine tip moth and the pitch pine tip moth. The subtropical pine tip moth's range closely follows the range of its primary host slash pine, extending through southern South Carolina south to southern Florida and west to Mississippi. The western pine tip moth has a wide distribution in the Western United States, from Montana, North and South Dakota, and Nebraska, to the Pacific Northwest, and south to Arizona and New Mexico.

Damage

Pine tip moths can reduce seedling growth, lead to stem deformity and can occasionally cause seedling mortality if infestation levels are high enough.

Diagnosis

Pine tip moth feeding damage in buds and shoots causes the affected plant tissue to turn brown (fig. 21.2). Dried resin can also be found on the buds and shoots where the insect bored into the seedling. Larval frass is evident when the damaged shoot or bud portion is dissected.

Biology

Pine tip moths have between one and six generations per year depending on moth species and climate, usually corresponding to the number of host growth flushes. Pine tip moths can overwinter as

Figure 21.2—*Nantucket pine tip moth damage on loblolly pine shoots.* Photo by Chris Asaro, Virginia Department of Forestry.

larvae or pupae, and usually overwinter in buds and shoots. Some species overwinter in the duff layer at the tree base. The adults emerge in spring and mate shortly after emergence. The female moths lay single eggs or clusters on shoots, needles, and buds (fig. 21.3). The newly hatched larvae mine needles before beginning to feed on buds and shoots underneath

Conifer Insects

21. Pine Tip Moths

Figure 21.3—*Pitch pine tip moth eggs. Note: darkened eggs have been parasitized by* Trichogramma *egg parasites.* Photo by Harry O. Yates, USDA Forest Service.

a protective web filled with pine resin (fig. 21.4). Larvae grow larger and begin to feed inside the buds and shoots, with the length of feeding damage somewhat proportional to the number of larvae feeding within each shoot.

Control

Chemical

A number of contact and systemic insecticides are available for tip moth. Contact insecticides can be quite effective at controlling tip moth infestations, but applications must be properly timed to reach newly hatched larvae before they enter buds and shoots. Typically, it is prohibitively expensive to spray each generation, especially over multiple years. Systemic insecticides are also effective and can provide 1 or 2 years of control from one application. They can only be applied at the seedling planting time or during the first year, however, and are ineffective if applied directly to larger trees.

Selected References

Asaro, C.; Fettig, C.J.; McCravy, K.M.; Nowak, J.T.; Berisford, C.W. 2003. The Nantucket pine tip moth: a literature review with management implications. Journal of Entomological Science. 38: 1–40.

Coulson, R.N.; Witter, J.A. 1984. Terminal, shoot, twig, and root insects. In: Forest entomology: ecology and management. New York: John Wiley & Sons: 445–456.

Figure 21.4—*Closeup of resin buildup on pine shoot with pupal exuvia from emerged tip moth.* Photo by Chris Asaro, Virginia Department of Forestry.

Forest Nursery Pests 83

Conifer Insects

22. Sawflies
John T. Nowak

Hosts and Distribution

Several sawfly species (families: Diprionidae and Tenthredinidae) are common pests of young conifers, including several introduced species. The more common species, hosts, and distribution are listed in table 22.1.

Damage

Although sawflies are common pests, they rarely cause seedling mortality. Defoliation is generally light, although localized epidemics have been reported. Sawflies are often divided into two groups: spring and summer sawflies. Spring sawflies generally feed on older foliage, and summer sawflies feed on both old and new foliage. The summer sawflies are the most destructive.

Diagnosis

Pine sawflies generally feed gregariously (fig. 22.1) in small groups and the larvae look similar to caterpillars with noticeable differences upon close inspection, including the number of prolegs (fig. 22.2) (sawflies have six or more pairs of prolegs and caterpillars have two to five pairs). Newly hatched larvae will often feed on needle edges. The damaged needles will turn brown and sometimes curl. As the larvae grow larger, they begin to consume the entire needle.

Table 22.1—*Common species, hosts, and distribution of sawflies in North America.*

Species	Hosts	Distribution
Redheaded pine sawfly *Neodiprion lecontei*	Scots, jack, red, shortleaf, loblolly, longleaf, and slash pine	Eastern United States and southeastern Canada
White pine sawfly *Neodiprion pinetum*	Eastern white pine	Throughout the range of its host
Introduced pine sawfly *Neodiprion similis*	Eastern white pine is preferred, also Scots, jack, and red pine	Eastern United States introduced
European pine sawfly *Neodiprion sertifer*	Scots, red, jack, and eastern white pine	Northeastern and Midwestern United States; parts of Ontario
Larch sawfly *Pristiphora erichsonii*	Tamarack and western larch	Great Lake States, eastern Canada, Washington, Oregon, Idaho, and Montana
Hemlock sawfly *Neodiprion tsugae*	Hemlock and Pacific silver fir	Coastal Oregon, Washington, and British Columbia; also interior forests of Montana, Idaho, and British Columbia
Lodgepole sawfly *Neodiprion burkei*	Lodgepole pine	Montana and Wyoming
Two-lined larch sawfly *Anoplonyx occidens*	Western larch	Northwestern United States
Yellowheaded spruce sawfly *Pikonema alaskensis*	Most species of spruce	Alaska, southern Canada, and Northern United States

Conifer Insects

22. Sawflies

Figure 22.1—*Pine sawfly damage on young pine sapling.* Photo by Albert (Bud) Mayfield, Florida Department of Agriculture and Consumer Services, at http://www.bugwood.org.

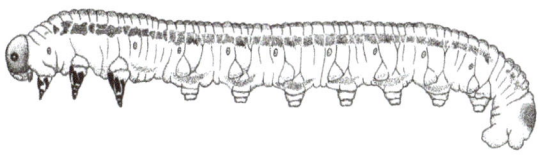

Figure 22.2—*Sawfly picture showing the typical number of prolegs (seven in this case).* Graphic by Randall Blackburn, Smithsonian Institution, at http://www.bugwood.org.

Biology

A generalized sawfly larvae is presented below. After the adults mate, the female adults (fig. 22.3) use their saw-like ovipositor to lay eggs in slits cut into needles. After feeding on the needles, the larvae will often drop to the forest floor and spin cocoons. For most conifer sawfly species, the last generation of the year will overwinter in a prepupal stage in the spun cocoon in the forest duff layer. Pupation occurs in spring. Some species overwinter in the egg stage. Spring sawflies generally have one generation per year, but summer sawflies often have multiple generations per year depending on climate.

Control

Cultural

For localized infestations, it may be practical to remove sawflies by hand or other means.

Chemical

Several contact insecticides are available for controlling sawfly larvae. Insecticide applications should target early stage larvae.

Conifer Insects

22. Sawflies

Figure 22.3—*Adult female sawfly laying eggs on needle with saw-like appendage.* Photo by Gyorgy Csoka, Hungary Forest Research Institute, at http://www.bugwood.org.

Selected References

Coulson, R.N.; Witter, J.A. 1984. Defoliating insects. In: Forest entomology: ecology and management. New York: John Wiley & Sons: 366–379.

Furniss, R.L.; Carolin, V.M. 1977. Western forest insects. USDA Forest Service Misc. Pub. 1339. Washington, DC: USDA Forest Service. 654 p.

Morris, C.L.; Hoffard, W.H. 1989. Sawflies. In: Cordell, C.E.; Anderson, R.L.; Hoffard, W.H.; Landis, T.D.; Smith, Jr., R.S.; Toko, H.V., tech. coords. Forest nursery pests. Agriculture Handbook 680. Washington, DC: USDA Forest Service: 82–83.

USDA Forest Service. 1985. Insects of eastern forests. USDA Forest Service Misc. Pub. 1426. Washington, DC: USDA Forest Service. 608 p.

Hardwood Diseases

Hardwood Diseases

23. Anthracnose
Jill D. Pokorny

Hosts

By definition, anthracnoses are leaf and twig diseases caused by a number of different but closely related fungi, all of which produce conidia in blister-like fruiting bodies called acervuli. Many anthracnose fungi are classified as ascomycetes and belong to genera including *Apiognominia, Gnomonia, Gnomoniella,* and *Glomerella*. The conidial states (anamorphs or asexual stages) are classified as coelomycetes and belong to genera including A*ureobasidium, Colletotrichum, Cryptocline, Diplodina, Discella,* and *Discula.*

Each anthracnose fungus is specific to the host tree it affects. For example, the oak anthracnose fungus infects oak trees only and will not spread to other tree species. Although these fungi are host-specific, several different anthracnose fungi can infect a single tree host. For example, several different anthracnose fungi can affect maple.

Anthracnose diseases have a very broad host range and affect many hardwood tree species in the United States. Symptoms are most severe on American sycamore, ash, maple, white oak, and black walnut. The disease is less common and damaging on hosts including birch, catalpa, elm, hickory, linden, bur, red and black oak, pecan, and yellow-poplar.

Distribution

Anthracnose diseases of hardwood seedlings occur throughout the range of the host species.

Damage

Anthracnose diseases may cause partial or complete defoliation of seedlings, resulting in a decrease in growth and vigor. Mortality of infected plants is rare. Anthracnose symptoms are more severe in years that have extended cool, wet spring weather.

Diagnosis

Symptoms vary with the tree host affected and the timing of the year the disease is observed. Symptoms are confined to the leaves on most tree hosts. However, on sycamore and oaks, the fungi may also affect buds, twigs, and shoots, and cause shoot blight, twig cankers, and branch dieback. On black walnut and hickory, the nuts may be affected.

Symptoms on infected leaves appear as dead areas (lesions) that range in size from small discrete spots (fig. 23.1) to medium size lesions (fig. 23.2) to large, irregular blotches (figs. 23.3 and 23.4). Lesions vary in color, and range from

Figure 23.1—*Small, discrete spots on maple.* Photo by Jill D. Pokorny, USDA Forest Service.

Figure 23.2—*Medium-sized lesions on black walnut.* Photo by Jill D. Pokorny, USDA Forest Service.

Figure 23.3—*Large, irregular leaf blotch lesions on white oak.* Photo by Joseph O'Brien, USDA Forest Service, at http://www.bugwood.org.

Hardwood Diseases

23. Anthracnose

Figure 23.4—*Large, irregular leaf blotch lesions on maple.* Photo by Jill D. Pokorny, USDA Forest Service.

Figure 23.5—*Vein-associated leaf blight lesion on sycamore.* Photo by Clemson University, USDA Cooperative Extension Slide Series, at http://www.bugwood.org.

black to reddish brown to tan. On sycamore (fig 23.5) and maple (fig. 23.6) trees, lesions often develop and extend along the leaf veins and midrib. Individual lesions may coalesce and cause large areas of the leaf to die, resulting in an overall wilted or scorched appearance to the leaves. Most infected leaves and leaflets fall prematurely.

When seedlings are infected early in the spring, the emerging leaves are often killed, turn black, and resemble damage caused by frost. If infection occurs during leaf expansion, growth of the infected tissue slows or stops as the rest of the leaf continues to expand. As a result of this unequal growth, the leaf tissue around the lesion becomes distorted and puckered (fig. 23.7). As leaves mature, they tend to become more resistant to infection and individual lesions are typically smaller in size.

Figure 23.6—*This lesion developed and expanded along a major leaf vein on maple.* Photo by Jill D. Pokorny, USDA Forest Service.

Figure 23.7—*Ash leaves exhibiting distorted and puckered growth adjacent to a large lesion.* Photo courtesy of the Plant Disease Clinic, University of Minnesota.

Forest Nursery Pests

Hardwood Diseases

23. Anthracnose

Blister-like fruiting bodies, called acervuli, form within the lesions on leaves, twigs, and fruit. They are particularly abundant on leaf veins and can be seen easily with a 10x hand lens (fig. 23.8). They vary in appearance from brown to pinkish in color. The pinkish color, when present, is that of conidia being produced in mass, within a mucilaginous matrix.

Biology

All fungi that cause anthracnose diseases have similar life cycles and require water from rain, dew, or fog to infect a tree. Because of these parameters, anthracnose is usually most severe in years with extended rainy weather periods in the spring and early summer. Disease development often subsides during summer months when environmental conditions become hot and dry.

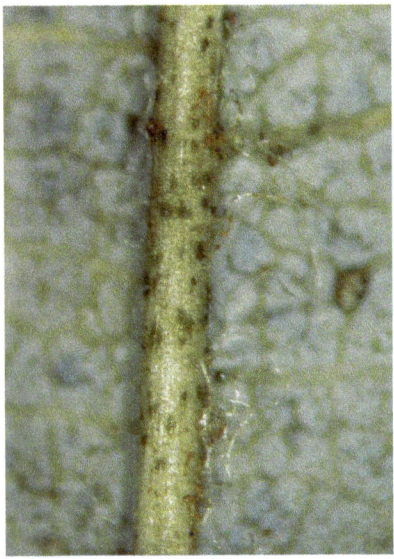

Figure 23.8—*Closeup of acervuli (brown dots) on and beside leaf veins on the lower leaf surface of a lesion on white oak.* Photo by Paul Bachi, University of Kentucky Research and Education Center, at http://www.bugwood.org.

Anthracnose fungi overwinter in infected leaf-and-twig debris on the ground or in cankered twigs and branches on the tree. During rainy periods in the spring, these fungi discharge large numbers of microscopic spores of the sexual state, called ascospores. The spores are spread by wind or splashing rain onto young, developing leaves of host seedlings. If the leaf surfaces remain wet for several hours, the spores germinate and produce lesions on the new leaves.

On most tree hosts, the fungi will produce secondary spores, called conidia, within the lesions that form on the leaves. Conidia are produced in large numbers and are also spread from leaf to leaf by wind and splashing rain. If rainy periods occur during the summer and fall, rapid increase and spread of anthracnose can occur by means of these secondary conidial spores. In some cases, such as maple anthracnoses, only conidial spores are present and the sexual states that produce ascospores, if any, are unknown.

Control

Prevention

Avoid this disease by planting anthracnose-resistant species or varieties when available. Some cultivars of London plane such as Columbia, Liberty, and the older Bloodgood, have a high level of resistance to anthracnose and are preferred. In the case of oaks, the red oak group is more resistant than the white oak group.

Cultural

Avoid close spacing and overhead irrigation of cuttings and young trees. Eliminate the overwintering fungus in plant materials in and around the nursery. Rake leaves and prune out severely infected twigs and branches to reduce the overwintering population of anthracnose fungi. Destroy infected material by burning or other appropriate means.

Chemical

Anthracnoses can be controlled with properly timed applications of a suitable fungicide. For effective control, fungicides must be applied before the disease appears in the spring. Apply the first spray at bud break, and repeat applications at 7- to 14-day intervals during cool, wet weather. Refer to the chemical labels for specific information.

To reduce chances for pathogens to develop chemical resistance, it is best to rotate the use of several different fungicides within the chemical spray program. Selected fungicides should have different modes of action and be applied in a rotational order.

Selected References

Berry, F.H. 1985. Anthracnose of eastern hardwoods. USDA Forest Service Forest Pest Leaflet 133. Washington, DC: USDA Forest Service. 8 p.

Berry, F.H. 1989. Anthracnose. In: Cordell, C.E.; Anderson, R.L.; Hoffard, W.H.; Landis, T.D.; Smith, Jr., R.S.; Toko, H.V., tech. coords. Forest nursery pests. Agriculture Handbook 680. Washington, DC: USDA Forest Service: 88–89.

Jones, R.K.; Benson, D.M. 2001. Diseases of woody ornamentals and trees in nurseries. St. Paul, MN: The American Phytopathological Society. 482 p.

Sinclair, W.A.; Lyon, H.H. 2005. Diseases of trees and shrubs, 2nd ed. Ithaca, NY: Cornell University Press. 660 p.

Wilson, D.; Carroll, G.C. 1994. Infection studies of *Discula quercina*, an endophyte of *Quercus garryanna*. Mycologia. 86: 635–647.

Hardwood Diseases

24. Leaf Spots and Blights

Michelle M. Cram and Will R. Littke

Hosts

Most hardwood species are affected by one or more fungi or bacteria that cause leaf spots and blights.

Distribution

Large varieties of fungi and bacteria cause leaf spots and often have wide distribution throughout their host's range.

Damage

The impact of leaf diseases on seedlings can range from minor loss of leaf area to entire crop loss. Leaf disease incidence can increase dramatically within nurseries due to seedling densities and frequent irrigation in nurseries. Defoliation of young seedlings in nurseries can lead to serious damage and mortality if the seedlings are unable to produce new leaves. Some leaf diseases will also infect the stem causing cankers that can result in stem breakage and mortality. Susceptibility to leaf diseases can vary greatly within a genus (for example, *Populus*). Damage by leaf diseases is typically reduced when resistant genotypes are used.

Diagnosis

The degree of leaf damage by a pathogen determines if it is characterized as a leaf spot or blight. Leaf spots are discreet necrotic areas a few millimeters to centimeters in size (fig. 24.1). Blight refers to a rapidly spreading leaf disease that kills most of the leaf, often including the stem, resulting in early leaf drop. Leaf spots vary in size, shape, and color depending on the pathogen and host. Most leaf diseases can be identified by the spores produced in fruiting bodies—typically black pycnidia or stromata.

Fungi in the *Phyllosticta* genus cause circular leaf spots that have a dark brown to purple outer ring and a light brown center with pycnidia eventually forming in the center (fig. 24.2). *Septoria* species produce a similar leaf spot (fig. 24.3). Other pathogens such as *Pseudocercospora* species tend to cause angular leaf spots by attacking the tissue between veins (fig. 24.4). Round, thick black structures called tar spot formed by *Rhytisma* species are common on maple and willow leaves (fig. 24.5). Another type of leaf disease caused by *Taphrina* species produces leaf blisters by stimulating infected cells to enlarge. These blisters begin as yellow bulges (2 to 20 mm) that last for a few weeks, eventually turning brown. Blisters can coalesce, affecting the entire leaf and resulting in defoliation (fig. 24.6).

Figure 24.1—*Tubakia leaf spot on oak caused by* Tubakia dryina. Photo by Michelle M. Cram, USDA Forest Service.

Figure 24.3—*Septoria leaf spot on red alder caused by* Septoria alnifolia. Photo by Will R. Littke, Weyerhaeuser Company.

Figure 24.2—*Leaf spot on red maple caused by* Phyllosticta minima. Photo by Michelle M. Cram, USDA Forest Service.

Figure 24.4—*Leaf spot on common persimmon caused by* Pseudocercospora fuliginosa. Photo by Michelle M. Cram, USDA Forest Service.

Hardwood Diseases

24. Leaf Spots and Blights

Figure 24.5—*Developing stromata of* Rhytisma punctatum *on bigleaf maple.* Photo by Will R. Littke, Weyerhaeuser Company.

Bacteria also cause leaf spots and blights and are common on trees in the *Prunus* genus (fig 24.7). Bacterial leaf spots appear as water-soaked dark brown or black spots that can enlarge during wet years leading to early defoliation. Bacterial leaf spots will ooze bacteria if cut and placed under a microscope.

Biology

Most leaf diseases have similar disease cycles. Sexual spores called ascospores are produced in leaves and stem tissue infected in the previous year. These spores are moved in the spring by wind and rain splash to infect new foliage. Further spread of the disease occurs during the asexual stage of the pathogen when conidia spores are produced in fruiting bodies throughout the late spring, summer, and early fall. These spores are spread by irrigation, rain splash, and wind to surrounding leaves, further intensifying the infection centers. Some pathogens of leaves also infect stems (for example, *Septoria* spp.) and cause cankers that result in stem breakage and mortality (see chapter 26).

Taphrina species do not produce spores in fruiting bodies; instead, ascospores are produced from a layer of asci that breaks through the surface of the infected leaf. The ascospores will then bud and form blastospores. These spores can infect newly expanding leaves or overwinter in bud scales and bark until bud break in the spring. Warm, wet springs favor the development of Taphrina leaf diseases, but serious infections are rare.

Bacteria that cause leaf diseases are spread by water and cutting tools and can infect leaves, twigs and fruit. Bacteria can overwinter in infected plant tissue and on plant surfaces until environmental conditions and host material are favorable for infection in the spring.

Figure 24.6—*Leaf blister on black cottonwood caused by* Taphrina populina. Photo by Will R. Littke, Weyerhaeuser Company.

Figure 24.7—*Bacterial leaf spot on chokecherry.* Photo by Michelle M. Cram, USDA Forest Service.

Hardwood Diseases

24. Leaf Spots and Blights

Control

Cultural

High seedling density and frequent irrigation favor leaf diseases. Culture techniques that encourage rapid drying of foliage can reduce the infection severity. Irrigate in the morning and consider lower seedling densities to permit more airflow and quick drying. After seedlings are lifted, incorporate all plant debris in the soil to reduce this inoculum source. In hardwood stooling beds, rogue any severely infected clones. Use resistant cultivars or clones if available.

Chemical

Control of leaf spots that affect seedling production often requires fungicides applied to the foliage every few weeks, beginning in the spring. Early population control of pathogens that cause leaf diseases is critical to avoiding premature defoliation. Pathogens that also cause stem cankers require regular inspection and control through the late summer.

Control of bacterial leaf spot diseases requires the use of copper-based fungicides and bactericides as preventative sprays. If copper-resistant strains of bacteria are present, it may be necessary to increase the action of copper by the addition of fungicides containing ethylene-bis-dithiocarbamate.

Selected References

Filer, Jr., T.H.; Anderson, R.L. 1989. Leaf spots and blights. In: Cordell, C.E.; Anderson, R.A.; Hoffard, W.H.; Landis, T.D.; Smith, Jr., R.S.; Toko, H.V., tech. coords. Forest nursery pests. Agriculture Handbook 680. Washington, DC: USDA Forest Service: 95–96.

Funk, A. 1985. Foliar fungi of western trees. Publication BC-X-265. Victoria, British Columbia, Canada: Canadian Forestry Service, Pacific Forest Research Centre. 159 p.

Scheck, H.J.; Pscheidt, J.W. 1998. Effect of copper bactericides on copper-resistant and –sensitive strains of *Pseudomonas syringae* pv. *syringae*. Plant Disease. 82: 397–406.

Sinclair, W.A.; Lyon, H.H. 2005. Diseases of trees and shrubs, 2nd ed. Ithaca, NY: Cornell University Press. 660 p.

Hardwood Diseases

25. Marssonina Blight

Michael E. Ostry and Jennifer Juzwik

Revised from chapter by Michael E. Ostry and Arthur L. Schipper, Jr., 1989.

Hosts

Marssonina blight affects all aspen and poplar species native to North America. The disease also damages many introduced poplar species and hybrids. Poplar species in the *Populus* section are susceptible to *M. brunnea* f. sp. *trepidae*, while species in the *Aigerios* section are susceptible to *M. brunnea* f. sp. *brunnea*. *Marssonina balsamiferae* has been reported occurring on balsam poplar in Ontario and Manitoba. *Marssonina castagnei* infects white poplar and *M. populi* is commonly found on quaking aspen.

Distribution

Four species of *Marssonina* and several formae speciales (f. sp.) are found throughout their poplar hosts' ranges. International transport of infected poplar cuttings may have contributed to the spread of the pathogens.

Damage

On highly susceptible poplar clones, the blight causes premature defoliation, significantly affecting growth. In addition to leaves and petioles, the fungi also infect young stems.

Diagnosis

Leaf symptoms vary, depending upon the susceptibility of the poplar species or hybrid, the species of *Marssonina*, and the severity of disease. Look for small, brownish, circular to angular spots, 1 to 2 mm across (fig. 25.1). These spots develop on leaves in spring and early summer. The center of the spots usually appears whitish. Spots may eventually join to form large, angular, rust-brown to black necrotic blotches (fig. 25.2).

In addition to leaf spots, lesions may develop in petioles (fig. 25.3) and succulent new stem growth (fig. 25.4). These lesions enlarge longitudinally and may become several millimeters in length. Whitish masses of conidia are produced from fruit bodies (acervuli) in the center of the lesions. Conidia of *Marssonina* species are oval, hyaline, and divided by one septum into a small basal cell and a larger upper cell. They often have one or more prominent vacuoles in each cell. The size of conidia varies with species: 11 to 21 by 4 to 7 microns (*M. brunnea*) (fig. 25.5), 15 to 23 by 5 to 8 microns (*M. castagnei*), 18 to 21 by 4.5 to 5.5 microns (*M. balsamiferae*) and 17 to 27 by 8 to 13 microns (*M. populi*). The perfect states of these fungi belong to the genus *Drepanopeziza* but are rarely seen.

Biology

Marssonina species overwinter in lesions on infected stems and on fallen leaves. In the spring during wet weather, the fungus releases ascospores (*Drepanopeziza* species) and possibly conidia, which were produced in fallen leaves or in lesions produced during the previous growing season. The spores are carried by wind and rain splash. Leaves, petioles,

Figure 25.1—*Leaf spots caused by* Marssonina brunnea *on a poplar leaf.* Photo by Michael E. Ostry, USDA Forest Service.

Figure 25.2—*Enlarged, angular leaf spots caused by* Marssonina brunnea. Photo by Michael E. Ostry, USDA Forest Service.

Figure 25.3—*Lesions caused by* Marssonina brunnea *on petioles of poplar leaves.* Photo by Michael E. Ostry, USDA Forest Service.

Hardwood Diseases

25. Marssonina Blight

Figure 25.4—*Lesions caused by* Marssonina brunnea *on poplar stems.* Photo by Michael E. Ostry, USDA Forest Service.

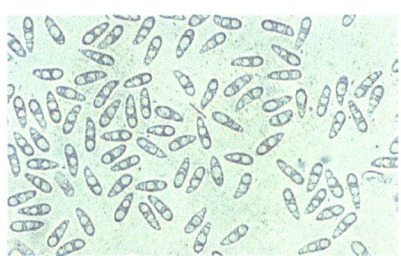

Figure 25.5—*Conidia of* Marssonina brunnea. Photo by Michael E. Ostry, USDA Forest Service.

and succulent stems of susceptible seedlings become infected, resulting in leaf spots and shoot lesions. Throughout the summer, conidia produced on these lesions are exuded in masses and disseminated by rain splash and wind to adjacent leaves and stems.

Control

Prevention

Several poplar species and hybrids resistant to *Marssonina* have been identified (fig. 25.6). Planting resistant clones and preventing the movement of these fungi on infected stock are the best disease preventative measures.

Cultural

Examine nursery stock for *Marssonina* infection. Cull infected hardwood cuttings to avoid spreading the fungus on planting stock. Remove and destroy, or plow under, diseased plant debris. Reduce planting density to improve air movement. When possible, avoid overhead irrigation systems.

Chemical

Fungicides can be used in the nursery to control Marssonina blight. Apply fungicides when disease symptoms first appear and make additional applications at recommended intervals. Rotate use of fungicides with different modes of action to reduce potential development of resistant pathogens.

Selected References

Newcombe, G.; Callan, B.E. 1997. First report of *Marssonina brunnea* f. sp. *brunnea* on hybrid poplar in the Pacific Northwest. Plant Disease. 81: 231.

Ostry, M.E. 1987. Biology of *Septoria musiva* and *Marssonina brunnea* in hybrid Populus plantations and control of Septoria canker in nurseries. European Journal of Forest Pathology. 17: 158–165.

Palmer, M.A.; Ostry, M.E.; Schipper, Jr., A.L. 1980. How to identify and control *Marssonina* leaf spot of poplars. St. Paul, MN: USDA Forest Service, North Central Forest Experiment Station.

Spiers, A.G. 1984. Comparative studies of host specificity and symptoms exhibited by poplars infected with *Marssonina brunnea*, *Marssonina castagnei*, and *Marssonina populi*. European Journal of Forest Pathology. 14: 202–218.

Spiers, A.G. 1988. Comparative studies of type and herbarium specimens of *Marssonina* species pathogenic to poplars. European Journal of Forest Pathology. 18: 140–156.

Spiers, A.G.; Hopcroft, D.H. 1998. Morphology of *Drepanopeziza* species pathogenic to poplars. Mycological Research. 102: 1025–1037.

Figure 25.6—*Leaves of poplar clones resistant (left) and susceptible to* Marssonina brunnea. Photo by Michael E. Ostry, USDA Forest Service.

Hardwood Diseases

26. Poplar Cankers
Michael E. Ostry and Jennifer Juzwik

Hosts and Causal Fungi

Bark necrosis, cankers, and dieback on native poplar species and hybrid poplars are caused by several species of fungi. The most common causal fungi are *Septoria musiva, Septoria populicola, Cytospora chrysosperma, Phomopsis macrospora, Dothichiza populea, Fusarium solani,* and *Lasiodiplodia theobromae*.

Distribution

These fungi are found throughout native and introduced hybrid poplars ranges in North America.

Damage

Poplar cankers can kill or severely weaken infected poplar seedlings and planted unrooted poplar cuttings. In addition, hardwood cuttings can be damaged during storage by a disease known as blackstem (fig. 26.1) caused by species of *Cytospora, Phomopsis,* and *Dothichiza*. Fungi introduced into plantations on infected nursery stock can result in planting failures (fig. 26.2).

Diagnosis

Symptoms resulting from stem infection by these fungi may be similar, particularly in early stages of the disease. Often it is necessary to examine spores or obtain cultures from diseased tissues to positively identify the causal fungus. In the spring, look for small necrotic or discolored areas with definite margins on stems. They are initially inconspicuous, but as the cankers develop, a distinct depressed area is observed in the affected bark. All of the possible pathogens, except *F. solani,* often produce pimple-like fruit bodies called pycnidia that protrude through the affected bark (fig. 26.3). White, yellow, or orange conidia, often seen as coils of spores, exude from the pycnidia during wet weather (fig. 26.4). When dry, these masses of conidia appear as fine tendrils. Some of the resulting cankers may girdle the stem, and the portion distal to the canker dies. *Septoria,* in addition to causing cankers (fig. 26.5), can cause leaf spots (fig. 26.6) and can prematurely defoliate highly susceptible clones.

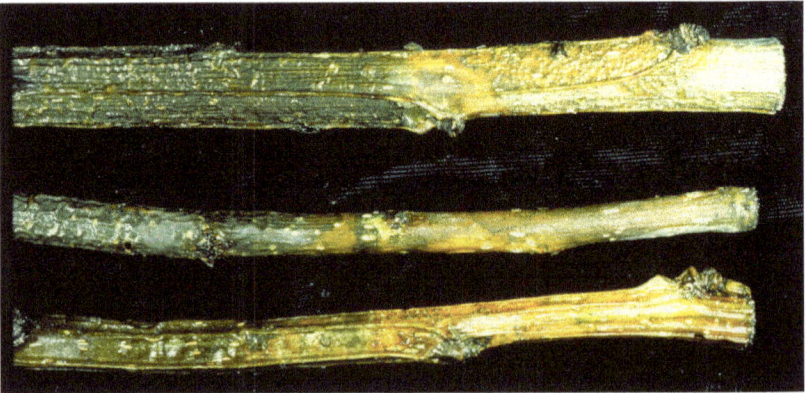

Figure 26.1—*Poplar cuttings affected by blackstem disease during improper storage.* Photo by Michael E. Ostry, USDA Forest Service.

Figure 26.2—*Planted poplar cutting killed by blackstem disease.* Photo by Michael E. Ostry, USDA Forest Service.

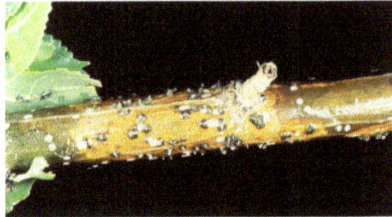

Figure 26.3—*Canker on poplar stem caused by* Dothichiza populea. Photo by Michael E. Ostry, USDA Forest Service.

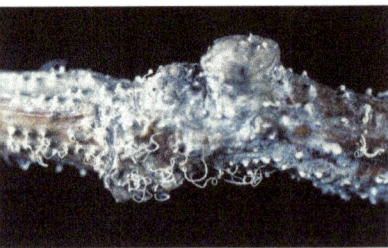

Figure 26.4—*Coiled tendrils of spores exuding from poplar stem infected by* Phomopsis macrospora. Photo by Michael E. Ostry, USDA Forest Service.

Hardwood Diseases

26. Poplar Cankers

Figure 26.5—*Canker caused by* Septoria musiva *on poplar stem*. Photo by Michael E. Ostry, USDA Forest Service.

Biology and Spread

These fungi overwinter in cankers on infected tree stems or on dead plant material on the ground. *Septoria* also overwinters in infected leaves. In the spring, the fungi produce spores that are spread by wind and rain splash. New infections occur during periods of high humidity that develop after overhead irrigation or rainfall throughout the summer.

Control

Prevention

Plant locally adapted, disease-resistant poplar clones to minimize damage. Removing native poplar trees near the nursery will reduce fungus inoculum. Many of these fungi colonize stressed trees so maintaining high tree vigor will minimize the risk of disease development.

Cultural

Use canker-free cuttings to establish nursery beds. Provide adequate water and nutrients to avoid tree stress. Infected poplar seedlings and unrooted poplar cuttings from diseased nursery stool beds should be culled to avoid shipment of diseased stock. Remove all leaves and debris after harvest to eliminate overwintering inoculum sources. Store cuttings at approximately -3 °C (26.6 °F) to avoid blackstem disease. Protect cuttings or rooted stock from drying out or becoming overheated during processing, storing, shipping, and planting.

Chemical

Fungicides can be used in the nursery to reduce some pathogen populations that lead to canker development. For example, control of the leaf disease *Septoria musiva* by fungicides can reduce disease pressure and canker development.

Selected References

Filer, Jr., T.H. 1967. Pathogenicity of *Cytospora*, *Phomopsis*, and *Hypomyces* on *Populus deltoides*. Phytopathology. 57: 978–980.

Filer, Jr., T.H. 1989. Poplar cankers. In: Cordell, C.E.; Anderson, R.L.; Hoffard, W.H.; Landis, T.D.; Smith, Jr., R.S.; Toko, H.V., tech. coords. Forest nursery pests. Agriculture Handbook 680. Washington, DC: USDA Forest Service: 101–102.

LeBoldus, J.M.; Blenis, P.V.; Thomas, B.R. 2008. Clone by isolate interaction in the hybrid poplar-*Septoria musiva* pathosystem. Canadian Journal of Forest Research. 38: 1888–1896.

Newcombe, G.; Ostry, M. 2001. Recessive resistance to Septoria stem canker of hybrid poplar. Phytopathology. 91: 1081–1084.

Ostry, M.E. 1987. Biology of *Septoria musiva* and *Marssonina brunnea* in hybrid Populus plantations and control of Septoria canker in nurseries. European Journal of Forest Pathology. 17: 158–165.

Ostry, M.E.; McNabb, Jr., H.S. 1982. How to identify and prevent injury to poplars caused by *Cytospora*, *Phomopsis*, and *Dothichiza*. St. Paul, MN: USDA Forest Service, North Central Forest Experiment Station. 8 p.

Ward, K.T.; Ostry, M.E. 2005. Variation in *Septoria musiva* and the implications for disease resistance screening of poplars. Plant Disease. 89: 1077–1082.

Figure 26.6—*Leaf spots on poplar leaves infected by* Septoria musiva. Photo by Michael E. Ostry, USDA Forest Service.

Hardwood Diseases

27. Poplar Leaf Rusts

Michael E. Ostry and Jennifer Juzwik

Hosts

All *Populus* species are potential hosts of poplar leaf rust fungi, *Melampsora* species. Distinguishing different leaf rust species is often problematic. Complexity in leaf rust situations is related to the extensive planting of many poplar species and interspecific hybrids, natural or human-aided movement of *Melampsora* species into new geographic regions, and the occurrence of *Melampsora* races (formae speciales). Furthermore, mixed race infections on single leaves and the evolution of natural hybrids of rust fungi contribute to this complexity. A critical need exists for monitoring poplar rust populations in this intricate system. Two poplar leaf rusts are dominant in the United States. In the East, larch is the alternate host of *M. medusae*; in the West, Douglas-fir is the alternate host for *M. occidentalis*. Infection of coniferous alternate hosts is confined to young needles and the damage is usually minor.

Distribution

Melampsora leaf rusts affect native and introduced hybrid poplars throughout North America.

Damage

Leaf rust can cause partial or complete defoliation by midsummer, reducing seedling vigor and quality. Premature defoliation can predispose seedlings to environmental stresses and invasion by secondary damaging agents.

Diagnosis

On poplar leaves, look for orange-yellow spore-bearing (uredinial) pustules (figs. 27.1 and 27.2). When these pustules

Figure 27.1—*Uredinia pustules with urediniospores of* Melampsora medusae *on poplar leaf.* Photo by Michael E. Ostry, USDA Forest Service.

Figure 27.2—*Uredinia pustules with urediniospores of* Melampsora occidentalis *on poplar leaf.* Photo by Michael E. Ostry, USDA Forest Service.

rupture, large numbers of powdery urediniospores are released. From late summer to autumn, yellowish crusts (telia), darkening to brown, are produced among the uredinia. On conifers, yellow spots occur on needles as the aecial stage develops. When mature, the aecia rupture and release powdery aeciospores. Aecia are preceded by an inconspicuous stage (pycnia) that produce their pycniospores in droplets (fig. 27.3). The uredinio-spores, teliospores, and aeciospores of *M. occidentalis* are generally larger than those of *M. medusae*.

Hardwood Diseases

27. Poplar Leaf Rusts

Figure 27.3—Melampsora medusae *pycniospore droplets on needles of larch.* Photo by Michael E. Ostry, USDA Forest Service.

Biology

Melampsora species require a coniferous alternate host to complete their life cycle. The conifer species varies with the rust species. Telia are produced on poplar leaves in the fall and appear as brown to black, crusty patches on fallen leaves. In spring, the teliospores in these fallen leaves germinate and produce basidiospores, which are windblown to the developing needles of coniferous hosts where infection takes place. The first fungal structures to develop on infected needles are the pycnia, which produce haploid spores (pycniospores) associated with sexual reproduction. In summer, shortly after the pycnia develop, aeciospores are produced on the needles and are windblown to poplar leaves. Yellow-orange uredinia pustules develop on infected poplar leaves and release urediniospores that are carried by wind to adjacent leaves, thus intensifying the disease on poplar hosts throughout the growing season. In the fall telia are produced, completing the annual disease cycle. Since both aeciospores and urediniospores are windborne, infection of poplars can occur at considerable distances from conifer hosts. In areas of the Southern United States where ornamental larch is grown, leaf rust epidemics can occur without need for windborne spores from larch in more northerly locations.

Control

Prevention

Planting poplar species and clones selected for resistance to the local *Melampsora* species population is the most effective practice. Growing poplars near coniferous alternate hosts will result in infection earlier in the season and increased disease severity. In the South, remove susceptible evergreen poplars to reduce the amount of overwintering uredinia. Leaf microflora and saprophytic microorganisms have been shown to be antagonistic to rusts and may function as natural biological control agents. Planting large, monoclonal beds should be avoided. This precautionary measure minimizes a new rust species or race from becoming damaging in the future.

Chemical

Application of fungicides labeled for control of Melampsora on poplar can reduce infection of leaves. Begin applying fungicidal sprays in the summer when conditions are favorable for spore development, dispersal, and infection.

Selected References

Innes, L.; Marchand, L.; Frey, P.; Bourassa, M.; Hamelin, R.C. 2004. First report of *Melampsora larici-populina* on *Populus* spp. in eastern North America. Plant Disease. 88: 85.

Moltzan, B.D.; Stack, R.W.; Mason, P.A.; Ostry, M.E. 1993. First report of *Melampsora occidentalis* on *Populus trichocarpa* in the Central United States. Plant Disease. 77: 953.

Newcombe, G.; Chastagner, G.A. 1993. First report of the Eurasian poplar leaf rust fungus *Melampsora larici-populina* in North America. Plant Disease. 77: 532–535.

Newcombe, G.; Chastagner, G.A. 1993. A leaf rust epidemic of hybrid poplar along the lower Columbia River caused by *Melampsora medusae*. Plant Disease. 77: 528–531.

Newcombe, G.; Stirling, B.; McDonald, S.; Bradshaw, Jr., H.D. 2000. *Melampsora xcolumbiana*, a natural hybrid of *M. medusae* and *M. occidentalis*. Mycological Research. 104: 261–274.

Ostry, M.E.; Wilson, L.F.; McNabb, Jr., H.S.; Moore, L.M. 1989. A guide to insect, disease, and animal pests of poplars. Agriculture Handbook 677. Washington, DC: USDA Forest Service. 118 p.

Shain, L. 1988. Evidence for *formae speciales* in the poplar leaf rust fungus, *Melampsora medusae*. Mycologia. 80: 729–732.

Shain, L.; Filer, Jr., T.H. 1989. Cottonwood leaf rusts. In: Cordell, C.E.; Anderson, R.L.; Hoffard, W.H.; Landis, T.D.; Smith, Jr., R.S.; Toko, H.V., tech. coords. Forest nursery pests. Agriculture Handbook 680. Washington, DC: USDA Forest Service: 90–91.

Stack, R.W.; Ostry, M.E. 1989. Melampsora leaf rust on *Populus* in the North Central United States. In: Merrill, W.; Ostry, M.E., eds. Proceedings, IUFRO Recent Research on Foliage Diseases. GTR-WO-56. Washington, DC: USDA Forest Service. 119–124.

Hardwood Diseases

28. Powdery Mildew

Michelle M. Cram and Glen R. Stanosz

Hosts

Powdery mildew diseases of hardwoods are primarily caused by fungi in the genera *Erysiphe*, *Phyllactinia*, *Pleochaeta*, and *Podosphaera*. Many of the fungi in these genera were previously classified as *Microsphaera*, *Sphaerotheca*, and *Uncinula*. Hardwood species commonly affected by powdery mildew include dogwood, yellow-poplar, oak, sycamore, cherry, maple, black walnut, hickory, buckeye, elm, and cottonwood.

Distribution

Powdery mildew diseases occur on hardwood seedlings grown throughout the United States. Worldwide, more than 500 fungi cause powdery mildew on 7,000 different plant species.

Damage

Infection of seedling leaves by powdery mildew can result in leaf distortion, chlorosis, reduced photosynthesis, and partial defoliation. Early infection and severe disease can result in reduced growth.

Diagnosis

White, powder-like colonies form on the surface of leaves and occasionally stems in the spring and early summer (figs. 28.1 and 28.2). In some species, these light-colored colonies can become brownish with age. Powdery mildew often affects the upper leaf, but can be found on the underside of the leaf as well. The powdery appearance of the disease is the result of conidia. These spores are produced in chains or singly during the asexual reproduction stage in the genera *Oidium*, *Ovulariopsis*, and *Streptopodium*.

Figure 28.1—*Flowering dogwood affected by a powdery mildew fungus.* Photo by Michelle M. Cram, USDA Forest Service.

The conidia are single-celled, colorless, and in the range of 20 to 50 by 12 to 20 microns. Diagnosis of these fungi to species is based on the individual characteristics of the conidiophores, conidia, and host.

Chasmothecia (syn. cleistothecia) are produced during the sexual reproduction stage on the surface of the plant, in or on the mycelia mat. Chasmothecia are spherical, typically 0.1 to 0.2 mm in diameter and change from colorless to yellow,

Hardwood Diseases

28. Powdery Mildew

Figure 28.2—*Powdery mildew fungus on Norway maple.* Photo by Glen R. Stanosz, University of Wisconsin-Madison.

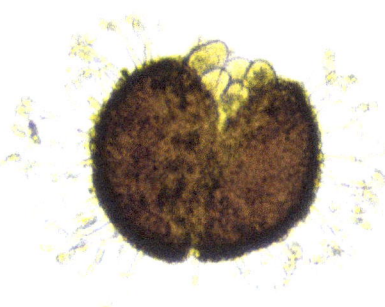

Figure 28.3—*Chasmothecium broken open to show asci and ascospores.* Photo by Glen R. Stanosz, University of Wisconsin-Madison.

then to brown, and finally to black. They have appendages to assist in anchorage and dispersal that are straight, flexuous, branched, or hooked. Chasmothecia and their appendages are barely visible with the naked eye, but seen more easily with a hand lens. Ascospores of the sexual stage are borne within one or more sac-like structures called asci (singular is ascus) that develop within the chasmothecia (fig. 28.3). Species identification is based on the characteristics of the chasmothecia including appendage type and ascus number and size.

Biology

Cool evenings and warm, humid days favor the development of powdery mildew. Typically, powdery mildew occurs when ascospores or conidia germinate, followed by infection of the host epidermal cells. Only a few species will parasitize the internal cells (mesophyll and palisade cells) of the host. Under favorable conditions of 15 to 28 °C (59 to 82 °F) new generations of conidia are produced every 4 to 6 days. These spores move by air and water to new infection sites and can germinate on dry surfaces. There are three patterns of the disease cycle. In warm regions, the fungi produce conidia throughout the year, creating multiple overlapping cycles of the disease. In temperate regions, these fungi survive winters and hot, dry summers by producing chasmothecia. As conditions become favorable again for disease development, ascospores are released to initiate new infections. A third type of disease cycle includes the overwintering of mycelium in the buds of woody plants or in mild climates on the leaves of evergreens. When the infected buds break and the new shoots develop, conidia are produced to initiate new infections.

Control

Cultural

Frequent irrigation and high seedling density enhances development of powdery mildew by maintaining high humidity over a longer period of time. Encourage rapid drying of foliage by irrigating in the morning and consider lower seedling densities. Because succulent tissues are highly susceptible, avoid over-fertilization with nitrogen. Reduce potential spring inoculum in the nursery by incorporating any infected plant debris into the soil after lifting.

Chemical

The use of fungicides during the spring and into the summer may be needed for powdery mildew control for some hardwood species. Early control of powdery mildew can be important to avoid damage and growth loss. Because host ranges of individual powdery mildew fungi vary considerably, a correct identification can help determine the need to protect other species being produced in the vicinity of diseased trees. Rotate fungicides with different modes of action to avoid development of resistance in powdery mildew fungi.

Selected References

Filer, Jr., T.H.; Affeltranger, C.E. 1989. Powdery mildews. In: Cordell, C.E.; Anderson, R.A.; Hoffard, W.H.; Landis, T.D.; Smith, Jr., R.S.; Toko, H.V., tech. coords. Forest nursery pests. Agriculture Handbook 680. Washington, DC: USDA Forest Service: 95–96.

Li, Y.; Mmbaga, M.T.; Windham, A.S.; Windham, M.T.; Trigiano, R.N. 2009. Powdery mildew of dogwoods: current status and future prospects. Plant Disease. 93: 1084–1092.

Sinclair, W.A.; Lyon, H.H. 2005. Diseases of trees and shrubs, 2nd ed. Ithaca, NY: Cornell University Press: 8–13.

102 Forest Nursery Pests

Hardwood Insects

Hardwood Insects

29. Cottonwood Borers
Forrest L. Oliveria and James D. Solomon

Hosts

Eastern cottonwood is the major host of the cottonwood borer (*Plectrodera scalator*) and the clearwing borer (*Paranthrene dollii*). Poplars and willows are also affected by these insects.

Distribution

The distribution corresponds closely with the eastern cottonwood range in the Eastern United States. The range of both borers extends westward into the Plains States; *P. scalator* is reported as far west as New Mexico and Montana. The largest populations occur in the Southern and Central States.

Damage

P. scalator larvae hollow, partially sever, or girdle seedling roots, causing structural weakening, loss of vigor, and mortality. Feeding by adult beetles of *P. scalator* often causes terminal death, followed by excessive branching, forking, and crooked stems. Stools for vegetative cutting production heavily infested with *P. dollii* do not produce vigorous shoots for vegetative cuttings. Some breakage occurs at tunneled sites.

Diagnosis

Initially, *P. scalator* infestations may go unnoticed because attacks occur at or below the groundline, and the larvae tunnel downward in the roots. As the infestation increases, look for plant weakening, mortality, and breakage near the groundline. Plants suspected of being infested should be lifted and have their roots examined. Infested roots are usually

Figure 29.1—*Galls on roots of cottonwood seedlings infested with* Plectrodera scalator. Photo by James D. Solomon, USDA Forest Service.

swollen and galled (fig. 29.1). They have breaks and openings in the bark, and frass often protrudes from gallery openings. In contrast, uninfested roots are comparatively smooth and uniform in shape.

Infested root dissection reveals galleries with one or more large, white, legless, longicorn-type larvae (fig. 29.2). Light brown, fibrous (usually excelsior-like) frass (fig. 29.3) is occasionally ejected from bark openings at the groundline.

From June through August, the large, black and white longhorn beetles (fig. 29.4) can be seen feeding on bark and terminals.

Attacks by *P. dollii* occur on the aboveground stem and are concentrated around the basal portion of the plant. Initial attacks are characterized by sap ooze and frass ejected from entrance holes. Attack sites often appear cankered and have enlarged entrances (fig. 29.5).

Figure 29.2—*Larva of* Plectrodera scalator *in root of cottonwood seedling*. Photo by James D. Solomon, USDA Forest Service, at http://www.bugwood.org.

Hardwood Insects

29. Cottonwood Borers

Figure 29.3—*Fibrous frass produced by larva of* Plectrodera scalator *at base of cottonwood seedling.* Photo by James D. Solomon, USDA Forest Service, at http://www.bugwood.org.

Figure 29.4—*Adult of* Plectrodera scalator. Photo by Charles T. Bryson, USDA Agricultural Research Service, at http://www.bugwood.org.

Infested stem dissection reveals galleries with one or more white to pinkish, caterpillar-like larvae with brown heads and thoracic shields (fig. 29.6). Granular frass piles (fig. 29.7), which are different in texture from the fibrous frass of *P. scalator*, often accumulate on the ground at the base of infested plants. Infested stems are commonly drilled by woodpeckers feeding on the larvae during winter.

P. dollii adults are dark, rusty red, clearwing moths (fig. 29.8) that closely mimic wasps.

Biology

Both borer species overwinter as larvae—*P. scalator* in roots and *P. dollii* in stools, trunks, and branches. *P. scalator* adults emerge mainly during June and July, cut niches in the bark, and lay eggs singly at or just below the groundline. Young larvae tunnel downward into the roots and produce galleries up to 25 mm (1.0 in) wide and 23 mm (0.9 in) long. The life cycle of *P. scalator* requires 1 to 2 years.

Figure 29.5—*Entrance holes in stem of cottonwood seedling, caused by larvae of* Paranthrene dollii. Photo by James D. Solomon, USDA Forest Service, at http://www.bugwood.org.

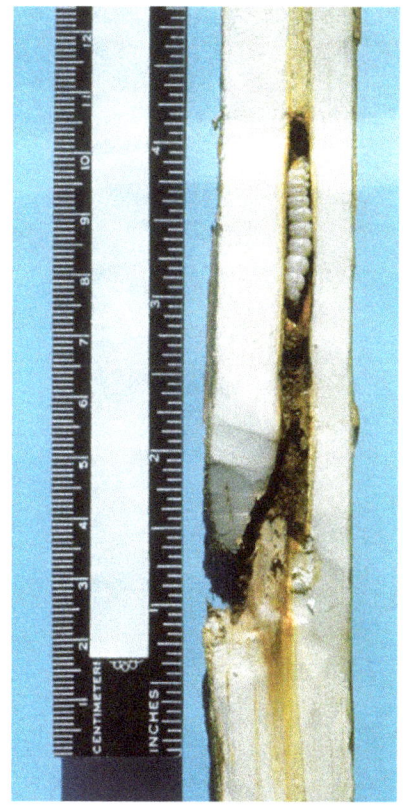

Figure 29.6—*Larva of* Paranthrene dollii *in stem of cottonwood seedling.* Photo by James D. Solomon, USDA Forest Service, at http://www.bugwood.org.

Hardwood Insects

29. Cottonwood Borers

Figure 29.7—*Granular frass of* Paranthrene dollii *at base of cottonwood seedling.* Photo by James D. Solomon, USDA Forest Service.

Figure 29.8—*Adult of* Paranthrene dollii. Photo by James D. Solomon, USDA Forest Service, at http://www.bugwood.org.

P. dollii has one generation per year. In the South, broods overlap, giving rise to moth emergence from April to November; in the North, moths emerge mostly during May and June.

Control

Prevention

Locate the nursery site 0.5 miles or more away from naturally occurring or planted poplars or willows to minimize insect invasion. Establish the nursery with uninfested cuttings or seedlings.

Cultural

Collect and promptly burn all branch, terminal, and basal trimmings and culled cuttings resulting from vegetative cutting operations to destroy hibernating insects.

Infested stools serve as the principal reinfestation reservoir for both *P. scalator* (in roots) and *P. dollii* (root collar). Therefore, dig and burn all sprout stools at 3-year intervals and replant with borer-free cuttings.

Chemical

The first application should be made 4 to 6 days after the first adults appear (about June 1 in Mississippi).

Selected References

Morris, R.C.; Filer, Jr., T.H.; Solomon, J.D. 1975. Insects and diseases of cottonwood. GTR-SO-8. New Orleans, LA: USDA Forest Service, Southern Forest Experiment Station. 37 p.

Solomon, J.D. 1980. Cottonwood borer (*Plectrodera scalator*)—a guide to its biology, damage, and control. RP-SO-157. New Orleans, LA: USDA Forest Service, Southern Forest Experiment Station. 10 p.

Solomon, J.D. 1989. Cottonwood borer. In: Cordell, C.E.; Anderson, R.L.; Hoffard, W.H.; Landis, T.D.; Smith, Jr., R.S.; Toko, H.V., tech. coords. Forest nursery pests. Agriculture Handbook 680. Washington, DC: USDA Forest Service: 106–108.

Hardwood Insects

30. Cottonwood Leaf Beetle
Forrest L. Oliveria and James D. Solomon

Hosts

Eastern cottonwood is the major host, particularly in the South, for the cottonwood leaf beetle (*Chrysomela scripta*). Poplars, willow, and alders are also affected.

Distribution

The cottonwood leaf beetle occurs throughout the United States but is most numerous in the lower Mississippi River Valley.

Damage

These beetles are serious defoliators of cottonwoods, particularly in the South and West. Continuing defoliation and twig damage through the summer reduces seedling growth and vigor. Lateral buds sprout below the injured terminals and grow rapidly, resulting in multiple-forked tops. Stunted growth in nursery plantings reduces cutting yield.

Diagnosis

Look for the sudden appearance of ragged foliage near branch ends and terminals (fig. 30.1). Some leaves will have brown patches where young larvae have skeletonized the leaves. Other leaves will have only their veins and midribs remaining. Heavy damage results in dead, black terminals with most of the leaf tissue consumed. Also look for black droppings on leaves.

Egg clusters, gregariously feeding larvae, and adult beetles are present on the affected foliage. The lemon-yellow eggs (fig. 30.2) are laid in clusters of 15 to 75 eggs on the underside of the leaves. Larvae are blackish to gray and about 12 mm long when mature (fig. 30.3).

Figure 30.1—*Damage to terminals and leaves of cottonwood caused by cottonwood leaf beetles.* Photo by James D. Solomon, USDA Forest Service, at http://www.bugwood.org.

Figure 30.2—*Lemon-yellow egg clusters of the cottonwood leaf beetle.* Photo by Whitney Cranshaw, Colorado State University, at http://www.bugwood.org.

Figure 30.3—*Larvae of the cottonwood leaf beetle.* Photo by James D. Solomon, USDA Forest Service, at http://www.bugwood.org.

Hardwood Insects

30. Cottonwood Leaf Beetle

Two whitish spots exist on the sides of each segment. When disturbed, the larvae release a pungent odor. Adults have a wide variety of color markings. Beetles are oval, yellow, and about 6 mm long, with slender black markings on their wingcovers (fig. 30.4). The head and thorax are black and the thorax margins are yellow or red.

Biology

The adults hibernate under bark, litter, and forest debris. They emerge in early spring and feed on unfolding leaves and tender buds at twig tips. In a few days, the female begins to lay eggs in clusters on the underside of leaves.

When the eggs hatch, the larvae begin to feed in groups on the underside of the foliage. Older larvae often feed separately and consume the entire leaf, except for the larger veins.

The pupae (fig. 30.5) attach themselves to leaves and bark or to weeds and grass beneath the trees.

Depending on latitude, two or more generations may develop per year. In Mississippi, a generation can develop in 35 days, and there may be up to seven generations per year.

The spring generation of the leaf-beetle may be greatly reduced by ladybird beetles, *Coleomegills maculate*, which feed on the eggs and pupae (fig. 30.6). Several other species of lady beetles, predaceous bugs, and two species of parasites also destroy leaf beetle eggs and larvae.

Control

Prevention

Use cottonwood clones that have demonstrated tolerance to leaf beetle defoliation.

Cultural

Employ sanitation practices in and around nurseries to either destroy the hibernating beetles directly or to expose them to winter temperatures.

Chemical

Schedule insecticide applications before larvae enter the pupal stage; treating seedlings at this time minimizes damage to predator populations.

Selected References

Morris, R.C.; Filer, Jr., T.H.; Solomon, J.D. 1975. Insects and disease of cottonwood. GTR-SO-8. New Orleans, LA: USDA Forest Service, Southern Forest Experiment Station. 37 p.

Neel, W.W.; Morris, R.C.; Head, R.B. 1976. Biology and natural control of the cottonwood leaf beetle, *Chrysomela scripta* (Fab.) (Coleoptera: Chrysomelidae). In: Thielges, B.A.; Land, Jr., S.B., eds. Proceedings, Symposium on eastern cottonwood and related species, September 28–October 2, 1976, Greenville, MS. Baton Rouge, LA: Louisiana State University: 264–271.

Oliveria, F.L.; Solomon, J.D. 1989. Cottonwood leaf beetle. In: Cordell, C.E.; Anderson, R.L.; Hoffard, W.H.; Landis, T.D.; Smith, Jr., R.S.; Toko, H.V., tech. coords. Forest nursery pests. Agriculture Handbook 680. Washington, DC: USDA Forest Service: 109–110.

USDA Forest Service. 1985. Insects of eastern forests. Misc. Publ. 1426. Washington, DC: USDA Forest Service. 608 p.

Figure 30.4—*Adult of the cottonwood leaf beetle.* Photo by Lacy L. Hyche, Auburn University, at http://www.bugwood.org.

Figure 30.5—*Pupae of the cottonwood leaf beetle.* Photo by James D. Solomon, USDA Forest Service, at http://www.bugwood.org.

Figure 30.6—*V-marked lady beetle feeding on cottonwood leaf beetle egg cluster.* Photo by James D. Solomon, USDA Forest Service, at http://www.bugwood.org.

Conifer and Hardwood Diseases

Conifer and Hardwood Diseases

31. Charcoal and Black Root Rots
Edward L. Barnard

Hosts

Charcoal root rot, caused by *Macrophomina phaseolina* (syn. *Sclerotium bataticola*), affects more than 300 plant species worldwide, including monocots, dicots, and gymnosperms. This disease has been considered one of the most serious diseases in conifer seedlings in forest tree nurseries in the Southern and Western United States. A closely related black root rot in southern forest nurseries has been attributed to *M. phaseolina* in combination with pathogenic *Fusarium* species and perhaps certain root parasitic nematodes.

Distribution

M. phaseolina and charcoal root rot occur in warm temperate and tropical regions of both the Eastern and Western Hemispheres. The pathogen is particularly prevalent in arid and semiarid environs. Black root rot has been described only in forest nurseries in the Southern United States, although *Fusarium* root rots and nematodes are common throughout North America (fig. 31.1).

Damage

Infections result in aboveground symptoms, ranging (and progressing) from damping-off of young seedlings, to stunted growth, to off-color wilting and reddening foliage, and eventual older seedling death (fig. 31.2). Infected seedlings are often irregularly distributed in nursery seedbeds, occurring singly or in clusters of a few to hundreds or thousands of seedlings, depending upon the distribution of infective inoculum in seedbed soils and associated predisposing stresses or cultural treatments (fig. 31.3). For example, it is not uncommon for infection foci to be concentrated at the ends of seedbeds or in seedbeds nearest to unfumigated irrigation pipelines where inoculum-laden soil may be mixed with fumigated soil during seedbed tilling and shaping operations. Also, charcoal root rot often flares up after stress-inducing cultural practices employed to improve

Figure 31.1—*Pine seedbeds in a Texas nursery severely affected by black root rot.* Photo by Edward L. Barnard, Florida Division of Forestry.

Figure 31.2—*Seedbeds with slash pine infected with* Macrophomina phaseolina. Photo by Edward L. Barnard, Florida Division of Forestry.

Figure 31.3—*Slash pine seedbeds infected with* Macrophomina phaseolina. Photo by Edward L. Barnard, Florida Division of Forestry.

seedling quality, such as undercutting, root wrenching, or withholding irrigation to harden-off seedlings for lifting.

Diagnosis

Root systems infected by *M. phaseolina* exhibit varying degrees of root discoloration and loss of fine feeder roots. Swelling, blackening, and cracking of infected cortical tissues may also occur, especially in advanced stages of disease development on taproots, larger lateral roots, and root collars (fig. 31.4). The presence of tiny black microsclerotia (50 to 200 microns) of *M. phaseolina* within and beneath cortical tissues of infected roots and root collars is common (fig. 31.5), especially where such tissues are moribund. *M. phaseolina* pycnidia have been observed on infected seedling stems, but these asexual spore-producing structures are far less common than microsclerotia. Microsclerotia and pycnidia may be visible to the unaided eye but are more readily observed with a hand lens.

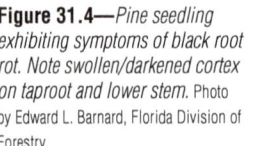

Figure 31.4—*Pine seedling exhibiting symptoms of black root rot. Note swollen/darkened cortex on taproot and lower stem.* Photo by Edward L. Barnard, Florida Division of Forestry.

Conifer and Hardwood Diseases

31. Charcoal and Black Root Rots

Figure 31.5—*Black microsclerotia of* Macrophomina phaseolina *on the surface of the xylem tissue beneath the bark of an infected slash pine seedling.* Photo by Edward L. Barnard, Florida Division of Forestry.

Biology

M. phaseolina is a high temperature, soil-borne, facultative parasite possessing a low competitive saprophytic ability; that is, it is easily suppressed by the activities of competitive, antagonistic, and hyperparasitic soil microflora. In culture, optimum temperatures between 30 and 37 °C (86 and 99 °F) are common for *M. phaseolina* isolates, and microsclerotia of the pathogen have been reported to survive soil temperatures in excess of 55 °C (131 °F). The fungus is capable of surviving in soils for years in the absence of a suitable host by means of resistant, resting state microsclerotia. However, survival of microsclerotia is strongly influenced by interactions of soil moisture and temperatures. Germination of microsclerotia is stimulated by a variety of organic substances, many of which may be supplied by host root exudates. Survival and activity of the pathogen in soils is responsive to inorganic fertilizers and organic soil amendments.

M. phaseolina is particularly aggressive on hosts subjected to various physiological stresses, particularly moisture stress. Thus, root disease episodes tend to peak late in the growing season when (1) soil temperatures approach their annual highs, (2) reduced rainfall and irrigation result in dry soils and sometimes excessively moisture-stressed seedlings, and (3) activity of beneficial soil microflora is restricted due to elevated soil temperatures and reduced soil moisture.

Control

Cultural

Several strategies are available to minimize the occurrence of charcoal and black root rots. Maintenance of adequate levels of soil organic matter (at least 2 percent) is helpful. Doing so promotes microbial activity, which may suppress *M. phaseolina*, an organism possessing low competitive saprophytic ability. Avoid excessive applications of nitrogen fertilizers in seedbeds, and favor nonhost cover crops such as millet over known hosts such as corn, peas, and sorghum. Avoid unnecessary, late season seedling stress because *M. phaseolina* is primarily aggressive on injured, water-stressed seedlings in warmer soils.

Chemical

Soil fumigation is recommended where charcoal root rot is persistent and problematic.

Selected References

Barnard, E.L. 1994. Nursery-to-field carryover and post-outplanting impact of *Macrophomina phaseolina* on loblolly pine on a cutover forest site in north central Florida. Tree Planters' Notes. 45(2): 68–71.

Barnard, E.L.; Gilly, S.P. 1986. Charcoal root rot of pines. Plant Pathol. Circ. 290. Gainesville: Florida Department of Agriculture and Consumer Services, Division of Plant Industry. 4 p.

Barnard, E.L.; Gilly, S.P.; Ash, E.C. 1994. An evaluation of dazomet and metam-sodium soil fumigants for control of *Macrophomina phaseolina* in a Florida forest nursery. Tree Planters' Notes. 45(3): 91–94.

Hodges, C.S. 1962. Black root rot of pine seedlings. Phytopathology. 53: 210–219.

Sinclair, W.A.; Lyon, H.H. 2005. Diseases of trees and shrubs, 2nd ed. Ithaca, NY: Cornell University Press. 660 p.

Smith, Jr., R.S.; Bega, R.V. 1964. *Macrophomina phaseolina* in the forest nurseries of California. Plant Disease Reporter. 48: 206.

Smith, Jr., R.S.; Hodges, C.S.; Cordell, C.E. 1989. Charcoal root rot and black root rot. In: Cordell, C.E.; Anderson, R.A.; Hoffard, W.H.; Landis, T.D.; Smith, Jr., R.S.; Toko, H.V., tech. coords. Forest nursery pests. Agriculture Handbook 680. Washington, DC: USDA Forest Service: 112–113.

Conifer and Hardwood Diseases

32. Cylindrocladium Diseases
Edward L. Barnard and Jennifer Juzwik

Hosts

Nearly 50 species of the closely related fungal genera *Cylindrocladium* and *Cylindrocladiella* induce diseases of various kinds on more than 130 genera of plants and trees worldwide. *Cylindrocladium* species including *C. scoparium*, *C. floridanum,* and *C. parasiticum* cause diseases on a wide variety of forest tree seedlings. Notable hosts in the North Central and Northeastern United States and Eastern Canada are red and eastern white pine and black, red, and white spruce. Highly susceptible hosts in the South include black walnut, yellow-poplar, sweetgum, eucalyptus, and eastern white pine. Other species that have been affected by *Cylindrocladium* species include cherrybark oak, northern red oak, flowering dogwood, and redbud seedlings.

Distribution

Cylindrocladium diseases of forest tree seedlings occur primarily in the North Central, Northeastern, and Southern United States. These diseases also occur in Ontario and Quebec, Canada. In addition, a species of *Cylindrocladium* has been isolated from soil samples in Washington State.

Damage

Cylindrocladium diseases, especially root infections, can cause significant seedling mortality (up to 80 percent) in forest nursery seedbeds and transplant beds (figs. 32.1 and 32.2). Sublethal infections may induce foliar chlorosis, stunting, or top dieback, resulting in costly culling of infected seedlings. Outplanting of nursery stock with undetected *Cylindrocladium* infections may also result in significant mortality within several years of plantation establishment.

Diagnosis

Depending on specific host, pathogen, and environmental combinations, *Cylindrocladium* species can cause pre- and post-emergence damping-off, foliage blight, stem lesions, and root rots. In northern conifers, lesions often accompanied with exuded resin masses may be visible on infected roots before any foliar symptoms are apparent. When the epidermis is removed from affected roots, reddish-brown cortical tissue can be seen. Advanced stage root rot on conifers results in lateral and primary root necrosis, often accompanied by blackening and slipping of the root cortical tissues (fig. 32.3). Root rot on hardwoods typically exhibits pronounced blackening, often with distinct longitudinal fissuring (fig. 32.4). Stem lesions on infected eucalyptus seedlings are frequently centered on leaf petioles, suggesting that infections beginning as leaf spots often progress to stems through the petioles. Foliage blights on black spruce and other conifers are characterized by foliar chlorosis, reddening and browning, necrosis, defoliation, and eventual seedling mortality in severe cases. Abundant asexual spores (conidia) may be produced on infected plant parts when conditions are suitable. Such fungal blooms often appear as a white, powdery, or granular covering or sheet.

Cylindrocladium species can be isolated directly from diseased tissue, indirectly from plant baits (for example, alfalfa seedlings and azalea leaves)

Figure 32.1—*Cylindrocladium root rot affected black spruce in a northern bareroot nursery field.* Photo by Jennifer Juzwik, USDA Forest Service.

Figure 32.2—*Bed of black walnut infected by Cylindrocladium species.* Photo by Charles E. Cordell, USDA Forest Service.

Conifer and Hardwood Diseases

32. Cylindrocladium Diseases

growing in or exposed to infested soil, or indirectly from soil when microsclerotia (groups of dark-colored, thick-walled cells) are captured on fine-mesh sieves. Once a *Cylindrocladium* has been isolated, the identity of the species present can be determined by microscopic observation of features of the asexual state. Distinguishing features are the shape of the terminal vesicle (swelling) of the conidiophores (spore-producing structures) and the size of the conidia (asexual spores). *C. parasiticum* and *C. floridanum* are characterized by spherical-shaped vesicles compared with the ellipsoidal to pear-shaped vesicles of *C. scoparium*. The straight conidia of *C. parasiticum* are larger (62 by 6 microns) than those of the straight conidia of *C. floridanum* (40.0 by 3.5 microns) and the straight to slightly curved conidia of *C. scoparium* (37 by 6 microns).

Figure 32.3—*Root infection of black spruce seedling by* Cylindrocladium floridanum. Photo by Jennifer Juzwik, USDA Forest Service.

Figure 32.4—*Symptomatic roots of yellow-poplar seedlings infected with* Cylindrocladium *species.* Photo by Charles E. Cordell, USDA Forest Service.

Conifer and Hardwood Diseases

32. Cylindrocladium Diseases

Biology

Cylindrocladium species survive and overwinter as microsclerotia in infected plant tissues and infested soils. The microsclerotia may persist in northern nursery soils for more than 15 years. Microsclerotia germinate and initiate new infections when stimulated by host root contact or exudates under suitable soil conditions. During periods of excessive moisture (humidity, rainfall, or overhead irrigation) and suitable temperatures, foliage and stem infections may develop from airborne asexual or sexual spores (conidia or ascospores, respectively). Perithecia (up to 1 mm) are reddish, enclosed structures of the sexual state that are not particularly abundant. They have been observed with *C. parasiticum* infections and on containerized longleaf pines infected with *C. floridanum* (sexual state *Calonectria kyotensis*) in Florida (fig. 32.5). *Cylindrocladium* species are apparently tolerant of a wide pH range, a reality that reduces the effectiveness of certain cultural control measures.

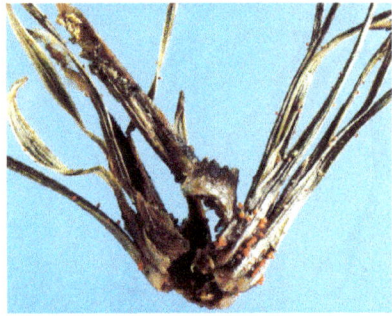

Figure 32.5—*Perithecia of* Calonectria kyotensis, *the sexual state of* Cylindrocladium floridanum, *on containerized longleaf pines.* Photo by Edward L. Barnard, Florida Division of Forestry.

Control

Cultural

Delineate and avoid infested nursery sites as much as possible. Restrict infected stock movement (for example, transplanting of susceptible stock within the nursery). Avoid spread of infectious propagules through movement of contaminated soil, equipment, or seedlings. Reduce disease spread by pressure washing tires and tillage implements between uses in Cylindrocladium disease-prone and other nursery fields. Favor nonhost cover crops such as corn and grasses (for example, sorghum-sudangrass) as opposed to known hosts such as soybeans, clover, and alfalfa in areas where *Cylindrocladium* has been problematic. Reduce the amount of organic substrate for *Cylindrocladium* population buildup by reducing sowing density and limiting cover crop development prior to soil incorporation. Maintain proper seedling spacing to promote aeration and growth while discouraging foliage and root infection. Rogue and cull badly damaged seedlings.

Chemical

Deep soil fumigation may be required for highly susceptible deep-rooted hardwoods including black walnut, yellow-poplar, and sweetgum. In northern nurseries, "preplant" soil fumigation is effective in dramatically reducing soil fungus populations and preventing the disease in subsequent crops. Judicious fungicide use may be helpful to prevent infections on certain conifer, hardwood, and eucalyptus species. Dipping transplant seedling roots in a fungicide solution is an effective control, but concerns regarding worker exposure to the chemical are significant.

Selected References

Barnard, E.L. 1984. Occurrence, impact, and fungicidal control of girdling stem cankers caused by *Cylindrocladium scoparium* on eucalyptus seedlings in a south Florida nursery. Plant Disease. 68: 471–473.

Cordell, C.E.; Barnard, E.L.; Filer, Jr., T.H. 1989. Cylindrocladium diseases. In: Cordell, C.E.; Anderson, R.A.; Hoffard, W.H.; Landis, T.D.; Smith, Jr., R.S.; Toko, H.V., tech. coords. Forest nursery pests. Agriculture Handbook 680. Washington, DC: USDA Forest Service: 114–117.

Crous, P.W. 2002. Taxonomy and pathology of *Cylindrocladium* (*Calonectria*) and allied genera. St. Paul, MN: The American Phytopathological Society (APS Press). 294 p.

Juzwik, J.; Honhart, C.; Chong, N. 1988. Cylindrocladium root rot in Ontario bareroot nurseries: estimate of losses in spruce seedlings. Canadian Journal of Forest Research. 18: 1493–1496.

Saunders, J.E.; Juzwik, J.; Hutchison, R. 1992. Outplanting survival of Cylindrocladium root rot affected black spruce seedlings. Canadian Journal of Forest Research. 22: 1204–1207.

Sinclair, W.A.; Lyon, H.H. 2005. Root rots and blights caused by *Cylindrocladium* and *Cylindrocladiella*. In: Diseases of trees and shrubs, 2nd ed. Ithaca, NY: Cornell University Press: 220–221.

Conifer and Hardwood Diseases

33. Damping-Off
Robert L. James

Hosts

Damping-off is a generic term that refers to the disease and mortality of recently germinated seeds and emerging seedlings. Several fungi and Oomycetes commonly cause damping-off in both bareroot and container nurseries, including *Fusarium, Rhizoctonia, Phytophthora,* and *Pythium* species. Most conifer and hardwood plant species are susceptible to damping-off, although junipers are an exception.

Distribution

Nurseries throughout tropical and temperate areas are prone to damping-off losses. Because damping-off is caused by a variety of pathogenic organisms with different environmental requirements, most nurseries experience some level of this disease.

Damage

Damping-off affects seeds and seedlings during germination and emergence. Pre-emergence damping-off occurs when germinating seeds are affected before their hypocotyls break through the soil surface. Post-emergence damping-off occurs shortly after seedling emergence when tissues are still succulent. Infection usually results in seed or seedling death.

Diagnosis

Losses from pre-emergence damping-off are often subtle and difficult to detect. Poor emergence in scattered pockets within bareroot beds or few or no emerging seedlings within containers, however, is a good indication that damping-off has occurred. Hardwood and conifer post-emergence damping-off symptoms are different. Post-emergence damping-off on hardwood seedlings appears as necrotic areas at or below the groundline. Infected seedlings wilt and die, but they often remain upright. Dead seedlings are brittle and break off easily just above the groundline. Post-emergence damping-off of conifers occurs between emergence and primary needle development. Infections also occur at or slightly below the groundline and result in water-soaked, discolored (usually brownish) areas that rapidly become sunken or constricted (fig. 33.1). Affected seedlings fall over (damp-off) (fig. 33.2). Seedlings that have fallen over are often green distal to the affected area until they dry out. The specific pathogen causing damping-off cannot be determined on the basis of symptoms. Identification usually requires infected tissue culturing and microscopic examination of associated organisms. Such identification is usually necessary before specific control recommendations can be made.

Biology

Most damping-off pathogens survive in soil, within plant debris, or on seeds (fig. 33.3), usually as viable dormant spores. They tend to increase when seedbeds are used continuously for seedling production without fallowing or preplanting soil fumigation. Damping-off organisms are not necessarily very aggressive pathogens. Pathogens may cause serious damage, however, when environmental conditions are conducive to their proliferation, especially when such conditions are detrimental to early seedling establishment. The principal environmental factors influencing disease development are soil pH, moisture, nutrition, and temperature; effects of these factors vary with the predominant pathogen and seedling species. All damping-off pathogens grow and reproduce best when the soil or growth medium pH is above the optimum for seedling growth. Cool, wet conditions slow germination and extend the time seeds and germinating seedlings are exposed to pathogens, increasing damping-off losses. Excessive soil moisture at moderate temperature favors *Pythium* and *Phytophthora* development. Soil texture, organic amendments, and nutrients can also influence disease severity. Fine-textured soils with high clay content retain water and warm

Figure 33.1—*Conifer seedling with typical foliar symptoms of post-emergence damping-off.* Photo by Robert L. James, USDA Forest Service.

Figure 33.2—*Newly emerged conifer seedling killed by post-emergence damping-off. Note discolored, constricted area of stem just above the groundline.* Photo by Robert L. James, USDA Forest Service.

Conifer and Hardwood Diseases

33. Damping-Off

Figure 33.3—*Fungal growth on seedcoat attached to an emerging seedling is a sign of damping-off.*
Photo by Robert L. James, USDA Forest Service.

slowly in the spring. Organic amendments may improve soil drainage, but may also provide substrates of potential damping-off pathogens. Nitrogen fertilizers applied when seedlings are emerging often increase seedling losses. Severe damping-off in bareroot nurseries is often associated with impeded surface or subsurface soil drainage. Damping-off in container nurseries is most often related to container reuse and seed contamination by potential pathogens.

Control

Cultural

Cultural practices can be modified to remedy conditions conducive to damping-off and can reduce potential damping-off organism inoculum. Soil pH can be lowered by adding sulfur compounds (granular sulfur or ammonium sulfate) or inorganic acids (sulfuric acid). Soil drainage can be improved in bareroot nurseries by leveling soil, installing subsurface drainage tiles, and adjusting irrigation frequency—which must be carefully controlled in container nurseries. Treating seeds and reused containers to reduce contamination by potential damping-off pathogens will greatly reduce losses. Also, minimize seed exposure to damping-off pathogens by delaying sowing until soil temperatures have risen to near optimum for rapid germination.

Chemical

Soil fumigation is the most effective way to control damping-off in bareroot nurseries. Soils should be fumigated when temperature, moisture, and physical condition are optimum for proper fumigant distribution. Pathogen-contaminated seeds should not be sown in fumigated soil because potential pathogen antagonists are usually eliminated by fumigation. Soil fungicide drenches are usually effective against damping-off in both bareroot and container nurseries. Pathogens causing disease, however, must be correctly identified because many modern fungicides are effective only against certain groups of related fungi.

Selected References

Hansen, E.M.; Myrold, D.D.; Hamm, P.B. 1990. Effects of soil fumigation and cover crops on potential pathogens, microbial activity, nitrogen availability, and seedling quality in conifer nurseries. Phytopathology. 80: 698–704.

Hartley, C. 1921. Damping-off in forest nurseries. Bull. 934. Washington, DC: USDA Bureau of Plant Industry. 79 p.

James, R.L. 1986. Diseases of conifer seedlings caused by seed-borne *Fusarium* species. In: Shearer, R.C., comp. Proceedings: Conifer Tree Seed in the Inland Mountain West Symposium. GTR-INT-203. Ogden, UT: USDA Forest Service, Intermountain Research Station: 267–271.

James, R.L. 1989. Effects of fumigation on soil pathogens and beneficial microorganisms. In: Landis, T.D., tech. coord. Proceedings: Intermountain Forest Nursery Association Meeting. GTR-RM-184. Fort Collins, CO: USDA Forest Service, Rocky Mountain Research Station: 29–34.

James, R.L.; Dumroese, R.K.; Wenny, D.L. 1990. Approaches to integrated pest management of Fusarium root disease in container-grown conifer seedlings. In: Rose, R.; Campbell, S.J.; Landis, T.D., eds. Target Seedling Symposium: Proceedings, Combined Meeting of the Western Forest Nursery Associations. GTR-RM-200. Fort Collins, CO: USDA Forest Service, Rocky Mountain Research Station: 240–246.

Kelley, W.D.; Oak, S.W. 1989. Damping-off. In: Cordell, C.E.; Anderson, R.L.; Hoffard, W.H.; Landis, T.D.; Smith, Jr., R.S.; Toko, H.V., tech. coords. Forest nursery pests. Agriculture Handbook 680. Washington, DC: USDA Forest Service: 118–119.

Landis, T.D.; Tinus, R.W.; McDonald, S.E.; Barnett, J.P. 1990. The container tree nursery manual. Vol. Five: The biological component: nursery pests and mycorrhizae. Agriculture Handbook 674. Washington, DC: USDA Forest Service. 171 p.

Rathbun-Gravatt, A. 1925. Direct inoculation of coniferous stems with damping-off fungi. Journal of Agricultural Research. 30: 327–339.

Tint, H. 1945. Studies in the Fusarium damping-off of conifers. II. Relation of age of host, pH, and some nutritional factors to the pathogenicity of Fusarium. Phytopathology. 35: 440–457.

Tinus, R.W.; McDonald, S.F. 1979. Greenhouse pest management. In: How to grow tree seedlings in containers in greenhouses. GTR-RM-60. Fort Collins, CO: USDA Forest Service, Rocky Mountain Forest and Range Experiment Station: 159–165.

Conifer and Hardwood Diseases

34. Fusarium Root and Stem Diseases
Robert L. James

Hosts

Fusarium diseases are caused by several *Fusarium* species, especially *F. oxysporum*, *F. proliferatum*, *F. solani*, and several species within the "*roseum*" complex, including *F. acuminatum*, *F. avenaceum*, *F. sambucinum*, *F. sporotrichioides*, and *F. equiseti*. Recent evidence indicates that some highly virulent strains previously identified as *F. oxysporum*, should be classified as *F. commune*. Although all conifer and many hardwood species may be affected, Douglas-fir, pines, larch, and true firs are most susceptible to Fusarium diseases. Spruce is often damaged less and species of cedar and cypress are relatively resistant to Fusarium diseases.

Distribution

Fusarium diseases are common throughout North America and in many temperate and tropical parts of the world. They especially have been a problem in Western Canada, and the North Central, Southern, and Western United States.

Damage

Fusarium species cause seedling root decay, stem cankers (including hypocotyl rot), seed decay, and wilt of some plant species within nurseries. Infected seedlings are often killed or have reduced growth and vigor. This pattern often results in increased numbers of culls and reduced survival after outplanting. Mortality is more common during the first growing season on bareroot stock and toward the end of the growth cycle on container seedlings. Losses can vary greatly depending on nursery management practices, seedling species, seedlot, geographic area, soil type, and environmental factors.

Diagnosis

Fusarium species cause a variety of disease symptoms within nurseries. Root disease usually results in primary and secondary root decay. Affected root systems often look black and lack actively growing root tips. The epidermis and cortex can often be easily stripped away from decayed roots. Aboveground root disease symptoms include: yellowing of foliage; needle tip and twig dieback; and foliar, branch, and stem necrosis (figs. 34.1, 34.2, and 34.3). Some *Fusarium* species also cause stem cankers either just above the groundline or higher on the main stem. Cankered seedlings often display aboveground symptoms similar to those of root-diseased seedlings. Fusarium sporulation can often be found on necrotic stem cankers or on seed coats and appear as orange to yellow pustules called sporodochia (fig. 34.4). *F. oxysporum* can also cause wilt of some nursery seedlings, resulting in foliage chlorosis and necrosis, and seedling tip and lateral branch bending. Root necrosis is not associated with wilt-causing *Fusarium*.

Analysis by a specialist is needed to identify *Fusarium* species. In a laboratory, *Fusarium* can be isolated from infected tissues or from soil using selective agar media and the different species identified by the unique fungal colony and reproductive morphology characteristics. In

Figure 34.1—*Severe mortality in bareroot seedlings due to Fusarium root disease.* Photo by Robert L. James, USDA Forest Service.

Figure 34.2—*True fir seedlings killed by Fusarium root disease. Note the random distribution of mortality.* Photo by Robert L. James, USDA Forest Service.

Conifer and Hardwood Diseases

34. Fusarium Root and Stem Diseases

Figure 34.3—*Containerized conifer seedlings with typical symptoms of Fusarium root disease.* Photo by Robert L. James, USDA Forest Service.

Figure 34.4—*Orange pustules of sporulating* Fusarium *species on an infected seed.* Photo by Robert L. James, USDA Forest Service.

Macrocondia are multicelled, slightly curved, and typically canoe-shaped (fig. 34.7). Microconidia usually have one or two cells, are ovoid or oblong, and borne in masses or in chains. Macroconidia and microconidia are formed within phialides on condiophores. Chlamydospores comprise the major resting structures of many *Fusarium* species. These chlamydospores have thick cell walls and form within plant substrates or macroconidia (fig. 34.8). A few *Fusarium* species, such as *F. oxysporum*, also produce long-lived resting structures called sclerotia. These sclerotia may remain viable within plant tissues or soil for many years. Light is required for some spore types to be produced.

Biology

Fusarium species have very wide host ranges, including many agricultural and natural plants and weeds. Host-specific strains of some pathogenic species occur, however. *Fusarium* strains can either be pathogenic (capable of causing plant disease symptoms) or saprophytic (not causing symptoms). Pathogenic and nonpathogenic strains are genetically distinct, but often appear morphologically identical. *Fusarium* species readily colonize nursery soil and organic matter and can remain dormant as chlamydospores and sclerotia in the absence of susceptible host plants. Diseased seedling roots from previous crops and organic amendments are major inoculum sources in bareroot nurseries. Reused containers and contaminated debris within greenhouse interiors are important inoculum sources

culture, the mycelium is extensive or appressed (pionnotal). Different *Fusarium* species produce different pigments in culture. For example, *F. oxysporum* and *F. proliferatum* usually produce dense white aerial mycelium and violet pigments best seen through the bottom of culture plates (fig. 34.5). Species in the "*roseum*" complex may produce carmine red to yellow mycelium and different intensities of carmine pigmentation (fig. 34.6). Four main spore types can be produced.

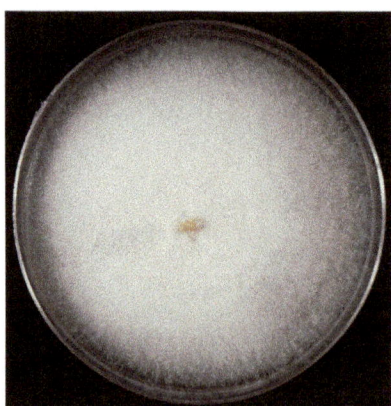

Figure 34.5—*Violet pigment produced by* Fusarium oxysporum *and* F. proliferatum *in culture.* Photo by Robert L. James, USDA Forest Service.

Figure 34.6—*Carmine pigment typical of species in the* Fusarium "*roseum*" *complex in culture.* Photo by Robert L. James, USDA Forest Service.

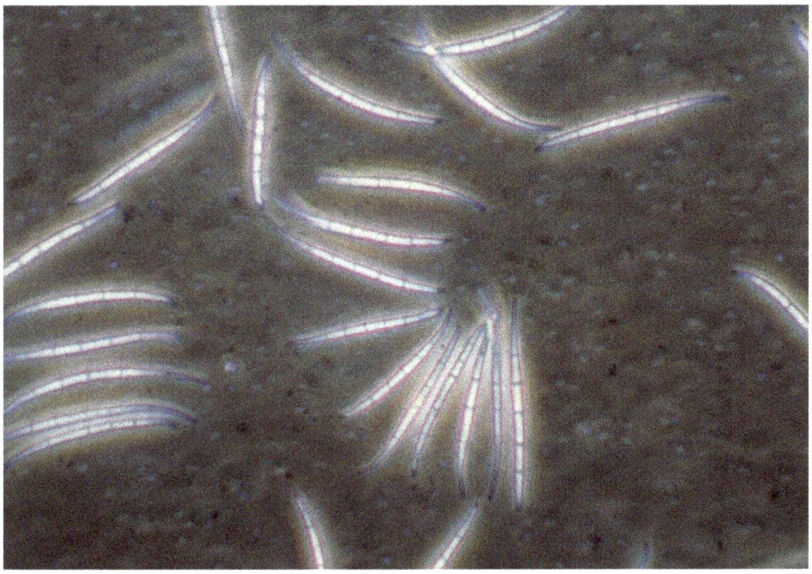

Conifer and Hardwood Diseases

34. Fusarium Root and Stem Diseases

Figure 34.7—*Multicelled, canoe-shaped macroconidia typical of* Fusarium *species. Details of shape and number of cells are among the characteristics used to identify species.* Photo by Robert L. James, USDA Forest Service.

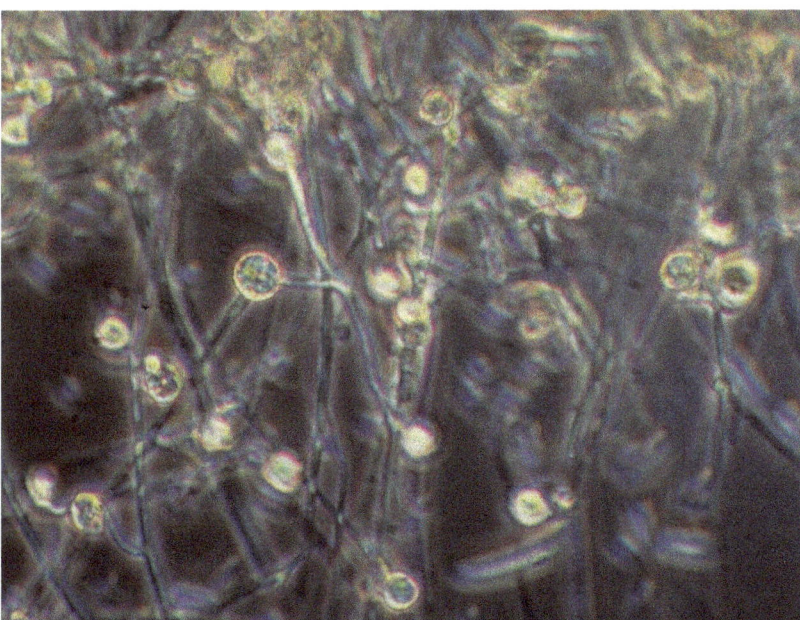

Figure 34.8—*Thick-walled chlamydospores can survive unfavorable conditions.* Photo by Robert L. James, USDA Forest Service.

in container operations. Many different plant seeds may be contaminated with *Fusarium* and may provide an important pathogen introduction source for nurseries. Chlamydospores germinate when root exudates of susceptible host plants are present and soil or greenhouse temperatures are warm. During warm, moist conditions, *Fusarium* grows on root surfaces, penetrates the epidermis, and spreads throughout root cortical tissues. Pathogenic strains penetrate host vascular systems, initiating disease symptoms. Some *Fusarium* species also produce toxins that aid host plant colonization and disease expression. Secondary disease spread can occur from aboveground inoculum (primarily macroconidia within sporodochia), especially on container stock within greenhouses. After the infected plants die and the fungus uses all available plant substrate nutrients, resting structures form and the cycle is repeated. Although some *Fusarium* species periodically form perfect (sexual) states, these spores are usually not as important in disease epidemiology as the more common asexual spores (macroconidia, microconidia, and chlamydospores).

Control

Prevention

The most effective way to reduce losses from Fusarium diseases is prevention, primarily by reducing pathogen inoculum levels within seedling growing environments. Prevention methods include treating seeds before sowing, adequately cleaning reused containers, cleaning greenhouse interiors between seedling crops, using pathogen-free growing media for container seedling production, and periodic diseased seedling

Conifer and Hardwood Diseases

34. Fusarium Root and Stem Diseases

removal during crop production cycles. Any seedling culls should not be added to soil for organic matter or stored near seed or transplant beds.

Cultural

Cover crops that control soil erosion and increase soil organic matter usually increase *Fusarium* populations after incorporation into the soil. The cover crop incorporated into the soil provides organic matter that stimulates pathogen population buildup. Leguminous cover crops are especially conducive to *Fusarium* level buildup. Fallowing fields for at least 1 year before sowing seeds helps reduce soilborne *Fusarium* populations. Some composts have potential to reduce *Fusarium* inoculum and subsequent seedling diseases. Avoiding seedling stress is important in reducing disease severity. It is important to maintain proper irrigation and balanced fertilization to avoid excess nitrogen levels.

Chemical

Preplant soil fumigation is the most common way to reduce *Fusarium* populations and subsequent disease in bareroot operations. Efficacy of particular fumigants varies among nurseries and is probably primarily related to soil characteristics including texture, density, and biological properties. Solarization (covering nursery soils with clear polyethylene tarps) is fairly effective in reducing *Fusarium* populations in some areas, but has limited efficacy on more northern sites. Commercial biological control agents (primarily antagonistic fungi, actinomycetes and bacteria) are available to help control Fusarium diseases. Their efficacy varies; most tests have resulted in disappointing disease control probably because biological control agents were not derived specifically from forest nurseries. Chemical fungicide application as a soil drench is usually ineffective after diseased seedling symptoms become evident, unless it is applied to reduce secondary disease spread from stem-colonizing pathogens, particularly within greenhouses.

Selected References

Brownell, K.H.; Schneider, R.W. 1983. Fusarium hypocotyl root rot of sugar pine in California forest nurseries. Plant Disease. 67: 105–107.

Gordon, T.R.; Martyn, R.D. 1977. The evolutionary biology of *Fusarium oxysporum*. Annual Review of Phytopathology. 35: 111–128.

Hamm, P.B.; Hansen, E.M. 1989. Fusarium hypocotyl rot. In: Cordell, C.E.; Anderson, R.L.; Hoffard, W.H.; Landis, T.D.; Smith, Jr., R.S.; Toko, H.V., tech. coords. Forest nursery pests. Agriculture Handbook 680. Washington, DC: USDA Forest Service: 120–121.

Hildebrand, D.M.; Stone, J.K.; James, R.L.; Frankel, S.J. 2004. Alternatives to preplant soil fumigation for western forest nurseries. PNW-GTR-608. Portland, OR: USDA Forest Service, Pacific Northwest Forest Research Station. 27 p.

James, R.L. 1986. Diseases of conifer seedlings caused by seed-borne *Fusarium* species. In: Shearer, R.C., comp. Proceedings: Conifer Tree Seed in the Inland Mountain West Symposium. GTR-INT-203. Ogden, UT: USDA Forest Service, Intermountain Research Station: 267–271.

James, R.L. 1997. A short review of *Fusarium* section *Liseola*: implications for conifer seedling production. In: James, R.L., ed. Proceedings of the Third Meeting of IUFRO Working Party S7.03-04 (Diseases and Insects in Forest Nurseries). Report 09-4: 34–41. Missoula, MT: USDA Forest Service, Northern Region, Forest Health Protection.

James, R.L.; Dumroese, R.K.; Gilligan, C.J.; Wenny, D.L. 1989. Pathogenicity of *Fusarium* isolates from Douglas-fir and container-grown seedlings. Idaho Forest, Wildlife and Range Experiment Station Bulletin. 52: 10.

James, R.L.; Dumroese, R.K.; Wenny, D.L. 1990. Approaches to integrated pest management of *Fusarium* root disease in container-grown conifer seedlings. In: Rose, R.; Campbell, S.J.; Landis, T.D., eds. Target Seedling Symposium: Proceedings, Combined Meeting of the Western Forest Nursery Associations. GTR-RM-200. Fort Collins, CO: USDA Forest Service, Rocky Mountain Research Station: 240–246.

James, R.L.; Dumroese, R.K.; Wenny, D.L. 1991. *Fusarium* diseases of conifer seedlings. In: Sutherland, J.R.; Glover, S.G., eds. Proceedings of the First Meeting of IUFRO Working Party S2.07-09 (Diseases and Insects in Forest Nurseries). Forestry Canada, Pacific and Yukon Region, Information Report BC-X-331: 181–190.

Johnson, D.W.; LaMadeleine, L.A.; Bloomberg, W.J. 1989. Fusarium root rot. In: Cordell, C.E.; Anderson, R.L.; Hoffard, W.H.; Landis, T.D.; Smith, Jr., R.S.; Toko, H.V., tech. coords. Forest nursery pests. Agriculture Handbook 680. Washington, DC: USDA Forest Service: 40–42.

Stewart, J.E.; Kim, M.S.; James, R.L.; Dumroese, R.K.; Klopfenstein, N.B. 2006. Molecular characterization of *Fusarium oxysporum* and *Fusarium commune* isolates from a conifer nursery. Phytopathology. 96: 1124–1133.

Conifer and Hardwood Diseases

35. Gray Mold
Diane L. Haase and Michael Taylor

Hosts

Gray mold is caused by the fungus *Botrytis cinerea*. Hundreds of woody and herbaceous species are affected. Spruce, redwood, giant sequoia, hemlock, larch, and Douglas-fir are all very susceptible. Container seedlings, bareroot seedlings grown at relatively high densities, and seedlings with some dieback (lower needle dieback or tissue damage caused by frost or other factors) are especially prone to infection.

Distribution

Botrytis cinerea is present in temperate regions worldwide.

Damage

Botrytis primarily affects needle tissue. Under heavy infection, it can penetrate stem tissues and kill the seedling. Roots tend to be unaffected while in the bareroot bed or in a container; however, after removal from the bed or container and placement in storage, roots can also be susceptible. Damage can occur in patches or can cause significant mortality in highly susceptible plants. Cold storage (at or above 0 °C, 32 °F) for longer than 2 weeks increases the risk of gray mold developing at damaging levels. *Botrytis* growth is inhibited in freezer storage (below 0 °C, 32 °F).

If less than 25 percent of the needles are damaged and no damage is observed on the stem, affected seedlings will often survive and grow well after outplanting where warmer, drier, aerated conditions are inhospitable to the mold. Seedlings with 25 to 75 percent needle loss and no stem damage may also survive in favorable outplanting environments but could have stunted growth in the first 1 to 2 years because of the reduced photosynthetic area. Seedlings with stem damage or a musty or pungent odor, regardless of the amount of gray mold present, are likely compromised and should be discarded.

Diagnosis

Symptoms usually occur in late summer and early fall but can arise year-round under favorable conditions in the nursery or in storage. *Botrytis* appears as a grayish-brown mold (mycelium), often on the lowermost foliage (fig. 35.1). Development of the mold leads to foliage discoloration and death (fig. 35.2). Microscopic fruiting bodies develop on infected seedlings and release puffs of spores when agitated (fig. 35.3). As the disease builds on dead needle tissue, it can spread into healthy shoots and into the stem, causing bark sloughing, cankering, and girdling.

Biology

Botrytis is a common saprophyte in forest nurseries. It can tolerate a wide temperature range (0 to 26 °C, 32 to 78 °F) with optimal development at 20 to 24 °C (68 to 75 °F). Inoculum usually originates from dead organic matter in or around the nursery and from senescent needles on the seedling shoot. *Botrytis* spores spread by air, water, and insects and can build to epidemic levels in a short time. The spores require free water to germinate. Under moist conditions, gray mold can spread vegetatively throughout the shoot

Figure 35.1—*Development of gray mold on lower foliage of stored Douglas-fir seedlings.* Photo by Diane L. Haase, USDA Forest Service.

Figure 35.2—*Extensive gray mold on foliage of containerized Engelmann spruce seedling.* Photo from USDA Forest Service Archive, at http://www.bugwood.org.

Conifer and Hardwood Diseases

35. Gray Mold

Figure 35.3—Botrytis cinerea *sporulating on Scots pine needles.* Photo by Petr Kapitola, State Phytosanitary Administration, Czechia, at http://www.bugwood.org.

and onto adjacent seedlings, especially with dense foliage conditions (such as tight spacing or after seedling canopy closure in late summer). In storage, dark, moist conditions and stagnant air further favor gray mold development.

Control

Cultural

Gray mold is best controlled by minimizing conditions that favor this disease. It is essential to keep the seedling canopy dry and well aerated. Humidity can be controlled in greenhouses with proper ventilation and spacing among plants. Irrigation can be applied in early morning to ensure foliage dries during the day. Nursery sanitation is critical. Removal of diseased seedlings and dead or dying organic matter (culls, weeds, etc.) minimizes sporulation of the pathogen. Storage units should be kept clean, cool, and well ventilated. In addition, minimizing cold storage duration and outplanting seedlots with incipient molding as soon as possible can greatly reduce or eliminate the impact of gray mold.

Chemical

Many fungicides are effective in gray mold control, most need direct contact with the affected material. Few are limited to use during specific life stages of the plant or fungus. It is important to rotate among at least three different chemical families to reduce the risk of the pathogen developing fungicide resistance. In greenhouses, applications should begin when the canopy closes and lower lying foliage reduces drying and air circulation, and should continue depending on species susceptibility or disease incidence until seedlings are hardened off. When applying to dense foliage, irrigation pressure and application rate should enable adequate canopy penetration. Application intervals should be 1 to 4 weeks, depending on species susceptibility or disease incidence. In bareroot nurseries, fungicides are applied as a gray mold preventative for highly susceptible species such as redwood or sequoia; otherwise, applications are made after the fungus is present, especially in high-density crops.

Biological

Application of beneficial bacteria (for example, *Streptomyces griseoviridis*) or fungi (for example, *Trichoderma* or *Gliocladium* species) have been shown to suppress incidence of *Botrytis* in forest nurseries.

Selected References

Capieau, K.; Stenlid, J.; Stenström, E. 2004. Potential for biological control of *Botrytis cinerea* in *Pinus sylvestris* seedlings. Scandinavian Journal of Forest Research. 19: 312–319.

Russell, K. 1990. Gray mold. In: Hamm, P.B.; Campbell, S.J.; Hansen, E.M., eds. Growing healthy seedlings: identification and management of pests in northwest forest nurseries. Portland, OR: USDA Forest Service, Forest Pest Management, Pacific Northwest Region, and Forest Research Laboratory, College of Forestry, Oregon State University: 10–13.

Srago, M.D.; McCain, A.H. 1989. Gray mold. In: Cordell, C.E.; Anderson, R.L.; Hoffard, W.H.; Landis, T.D.; Smith, Jr., R.S.; Toko, H.V., tech. coords. Forest nursery pests. Agriculture Handbook 680. Washington, DC: USDA Forest Service: 45–46.

Williamson, B.; Tudzynski, B.; Tudzynski, P.; Van Kan, J.A.L. 2007. *Botrytis cinerea*: the cause of grey mould disease. Molecular Plant Pathology. 8: 561–580.

Conifer and Hardwood Diseases

36. Nematodes
Stephen W. Fraedrich and Michelle M. Cram

Hosts

Seedlings of most tree species are susceptible to damage caused by plant parasitic nematodes. Some plant parasitic nematodes have wide host ranges that may include woody plants, cover crops, and weeds. Other nematodes favor only certain plant species. A large number of nematode species have been found to damage forest-tree seedling crops. Among the most damaging nematodes are the stunt nematodes (*Tylenchorhynchus* spp.), dagger nematodes (*Xiphinema* spp.), needle nematodes (*Longidorus* spp.), root-knot nematodes (*Meloidogyne* spp.), pine cystoid nematodes (*Meloidodera* spp.), lance nematodes (*Hoplolaimus* spp.), stubby-root nematodes (*Trichodorus* spp.), and lesion nematodes (*Pratylenchus* spp.).

Distribution

Plant parasitic nematodes can affect forest-tree seedling production in all areas of the United States. More nematode species, however, occur in warmer regions, and seedlings produced in the South are more likely to be affected. Some nurseries in the Midwestern and Pacific Northwestern United States and Western Canada have reported problems with several nematode species, but documented cases of damage are rare in nurseries located in other areas of the western and northern regions of the United States.

Damage

Plant parasitic nematodes feed directly on seedling roots resulting in damage to the root system. Nematode feeding can reduce nutrient uptake, suppress root elongation, and generally interfere with the normal metabolic processes of seedlings. Wounds created by nematodes can also serve as infection courts for fungal pathogens that further debilitate roots. Seedlings affected by nematodes lack vigor and are usually stunted, although even severely stunted seedlings will often survive in seedbeds until they are challenged with a period of soil moisture stress. Stunting caused by nematodes reduces seedling quality and can render seedlings unsuitable as planting stock. Seedlings are particularly vulnerable to damage by nematodes during the weeks that follow seed germination. Later in the growing season, larger seedlings can better withstand some nematode feeding without noticeable effects.

Diagnosis

Nematode feeding produces various symptoms in seedlings that can be helpful in alerting nursery managers to problems with nematodes, but correct diagnosis requires nematode extraction from soil and roots. Seedlings damaged by nematodes exhibit stunting, and foliage may be reduced in size and appear chlorotic (fig. 36.1). Seedling stunting may not be uniform throughout fields and can give fields a wavy appearance (figs. 36.1 and 36.2). The root systems of nematode-damaged seedlings are often greatly reduced in size compared with

Figure 36.1—*Chlorosis and stunting in slash pine seedlings caused by the stunt nematode* (Tylenchorhynchus claytoni). Photo by Michelle M. Cram, USDA Forest Service.

Figure 36.2—*Stunting in loblolly pine seedlings caused by the needle nematode* (Longidorus americanus). Photo by Stephen W. Fraedrich, USDA Forest Service.

Conifer and Hardwood Diseases

36. Nematodes

undamaged, healthy seedlings (fig. 36.3). Nematode species such as root-knot or dagger nematodes may induce galls on roots, but most nematodes produce no such definitive symptoms. Less obvious symptoms produced by nematodes can include swelling of roots, lesions, or a reduction in root elongation. Wounds caused by nematodes may provide an entry court for fungal pathogens resulting in root rot and seedling wilting.

Cooperative extension offices, some Federal agencies, and private companies evaluate soil and plant samples for plant parasitic nematodes. A variety of techniques should be used to account for ectoparasitic and endoparasitic nematodes. Roots should be examined and sampled for endoparasitic nematodes, such as lance nematodes, and sedentary nematodes, such as root-knot nematodes. Soil extraction techniques with the Baermann funnel or the centrifugal flotation methods are often adequate for most ectoparasitic nematodes that are less than 2 mm in length. These extraction techniques may have to be modified for larger nematodes, such as needle nematodes, by using wider mesh screens in the Baermann funnel method or higher sugar concentrations in the centrifugal flotation method.

Biology

Most plant parasitic nematodes are worm-like for a least a part of their life cycle. Their lengths generally range from 0.5 to 2.0 mm, although some species, such as those in the genus *Longidorus,* can be up to 11 mm in length. Nematodes with sedentary stages, in which females attach to roots and swell as they feed, ultimately have a round to pear-shaped appearance. Plant parasitic nematodes possess a hollow stylet, which is used to puncture root cell walls and suck the contents from cells (fig. 36.4). Plant parasitic nematodes are either ectoparasites that feed outside the roots or endoparasites that burrow inside the roots to feed. Nematodes have an egg stage, and most go through four juvenile stages before reaching the adult stage. Usually, nematodes can complete their life cycle in just 3 to 6 weeks under optimum conditions, although some nematode species require 1 year or more to complete a life cycle. Nematodes are most common in the upper 30 cm (12 in) of soil but can occur deeper in soil. Some nematode species can also migrate vertically through the soil in response to temperature and moisture fluctuations.

Control

Prevention

Movement of plant parasitic nematodes by equipment is a primary way that infestations spread. After a field is identified as infested, restrict equipment movement through the field or wash any nematode-infested soil and debris off the equipment before moving to uninfested fields.

Figure 36.3—*Healthy loblolly pine seedlings (left), and seedlings exhibiting slight (center) and severe (right) stunting caused by the stunt nematode* (Tylenchorhynchus ewingi). Photo by Stephen W. Fraedrich, USDA Forest Service.

Conifer and Hardwood Diseases

36. Nematodes

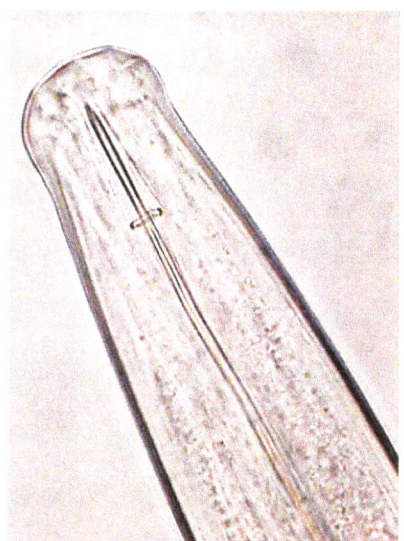

Figure 36.4—*Needle nematode* (Longidorus americanus) *head with long stylet.* Photo by Michelle M. Cram, USDA Forest Service.

Cultural

Crop rotations with nonhost crops can prevent the buildup of plant parasitic nematodes in nurseries. For instance, the needle nematode, *L. americanus*, feeds and reproduces on pines and oaks but does not feed on grasses such as sorghum-sudan and rye. In contrast, the stunt nematodes have wide host ranges that include pines, grasses, and legumes. Pearl millet varieties have been identified as nonhosts or poor hosts for many different plant parasitic nematodes, including species of root-knot, lesion, sting, and needle, and stunt nematodes. Information about the host range for individual nematode species is necessary to effectively use crop rotation.

The use of fallow can be an effective strategy for controlling many nematode species. Nematodes require host plants for survival in fields, and maintaining fallow fields can starve nematodes. This technique requires controlling weeds that might be hosts for the nematodes. Periodic soil disking or tilling can also hasten the decline of nematode populations in fields. Nematodes that can live in soil for extended periods without hosts, such as the dagger nematode *X. diversicaudatum*, are unlikely to be controlled by fallow or crop rotations and may require chemical controls.

Chemical

Broad-spectrum fumigants provide excellent control of plant parasitic nematodes. Fumigants are typically applied to the upper 15 to 20 cm (6 to 8 in) of soil and can significantly reduce nematode populations during the first year. However, fumigation may not eliminate nematodes in deeper soil layers, soil clods, or in roots not penetrated by fumigants. In addition, nematodes can be easily moved from infested fields to fumigated fields by equipment. Thus, nematode populations often rebound on seedling crops after fumigation, making it necessary to refumigate fields, rotate with nonhosts, or fallow before the next seedling crop.

Selected References

Cram, M.M; Fraedrich, S.W. 2005. Management options for control of a stunt and needle nematode in southern forest nurseries. In: Dumroese, R.K.; Riley, L.E; Landis, T.D. RMRS-P-35. National Proceedings: Forest and Conservation Nursery Associations—2004. Fort Collins, CO: USDA Forest Service, Rocky Mountain Research Station: 46–50.

Cram, M.M.; Fraedrich, S.W. 2009. Stunt nematode (*Tylenchorhynchus claytoni*) impact on southern pine seedlings and response to a field test of cover crops. In: Dumroese, R.K.; Riley, L.E. RMRS-P-58. National Proceedings: Forest and Conservation Nursery Associations, 2008. Fort Collins, CO: USDA Forest Service, Rocky Mountain Research Station: 95–100.

Ruehle, J.L. 1973. Nematodes and forest trees—types of damage to tree roots. Annual Review of Phytopathology. 11: 99–118.

Ruehle, J.L.; Riffle, J.W. 1989. Nematodes. In: Cordell, C.E.; Anderson, R.A.; Hoffard, W.H.; Landis, T.D.; Smith, Jr., R.S.; Toko, H.V., tech. coords. Forest nursery pests. Agriculture Handbook 680. Washington, DC: USDA Forest Service: 122–123.

Shurtleff, M.C.; Averre, III, C.W. 2000. Diagnosing plant disease caused by nematodes. St. Paul, MN: APS Press: 189 p.

Sinclair, W.A.; Lyon, H.H. 2005. Diseases of trees and shrubs, 2nd ed. Ithaca, NY: Cornell University. 660 p.

Sutherland, J.R.; Sluggett, L.J.; Lock, W. 1972. Corky root disease observed on two spruce species and western hemlock. Tree Planters' Notes. 23(4): 18–20.

Timper, P.; Hanna, W.W. 2005. Reproduction of *Belonolaimus longicaudatus*, *Meloidogyne javanica*, *Paratrichodorus minor*, and *Pratylenchus brachyurus* on Pearl Millet (*Pennisetum glaucum*). Journal of Nematology. 37: 214–219.

Whitehead, A.G. 1998. Plant nematode control. New York: CABI Publishing. 384 p.

Conifer and Hardwood Diseases

37. Phytophthora Root Rot
Michelle M. Cram and Everett M. Hansen

Hosts

Phytophthora root rot occurs on a broad range of conifers and hardwoods. Tree species most often affected by *Phytophthora* in forest nurseries include black walnut, Fraser fir, Douglas-fir, and red pine. Other *Phytophthora* hosts in forest nurseries include species of true fir, hemlock, larch, pine, spruce, white-cedar, yew, oaks, birch, and American chestnut. Species identified as highly tolerant of *Phytophthora* include western redcedar, western larch, lodgepole pine, ponderosa pine, and Chinese chestnut.

Distribution

Phytophthora root rot is predominately a problem in bareroot forest nurseries throughout the United States. The disease develops in nurseries where susceptible seedlings are grown in saturated soils. Damage by *Phytophthora* in bareroot forest nurseries occurs most often in low wet areas or poorly drained areas (fig. 37.1). Damage in container nurseries occurs when container soil becomes contaminated either by storing containers and potting soil on infested ground or by irrigation from infested water sources. Some common *Phytophthora* species that cause root rot and stem damage include *P. cinnamomi*, *P. citricola*, *P. cactorum,* and *P. drechsleri*. In the Pacific Northwest conifer seedlings are also affected by *P. sansomeana*, *P. megasperma*, *P. pseudotsugae*, *P. cryptogea*, and *P. gonapodyides*.

Damage

Phytophthora root rot can extend to the root collars and stems of susceptible conifer and hardwood seedlings, affecting seedling quality and mortality. Seedlings lifted with an incipient *Phytophthora* infection on the root systems can result in seedling losses in storage, transport, and outplanting. Whole nursery fields of highly susceptible hosts have been destroyed due to infestation by *Phytophthora*.

Figure 37.1—*Phytophthora root rot develops on seedlings in low wet and poorly drained areas.* Photo by Everett M. Hansen, Oregon State University.

Diagnosis

General aboveground symptoms of Phytophthora root rot on seedlings include chlorosis, stunting, wilting, and eventual mortality (fig. 37.2). The disease usually occurs in patches as infection spreads from seedling to seedling. The cambium tissue of infected conifers turns a reddish-brown or butterscotch color at the root collar and in the primary and lateral roots (fig. 37.3). Fine feeder roots infected by *Phytophthora* are often black, decayed, or missing.

Black walnut is the hardwood species most susceptible to Phytophthora root rot. Black walnut infected tissue becomes black, soft, and water-soaked at the root collar, extending into the stem and primary root (fig. 37.4). Oak and chestnut seedlings have similar damage patterns, but the infected tissue is brown or tan.

Positive identification of *Phytophthora* infection requires using the services of a pathologist or commercially available test kits. Diagnostic kits for *Phytophthora* provide a quick identification in the field.

Figure 37.2—*Symptoms of Phytophthora root rot in a black walnut bed.* Photo by Michelle M. Cram, USDA Forest Service.

Conifer and Hardwood Diseases

37. Phytophthora Root Rot

Figure 37.3—*Red-brown or butterscotch colored cambium tissue of a Douglas-fir seedling infected by* Phytophthora. Photo by Everett M. Hansen, Oregon State University.

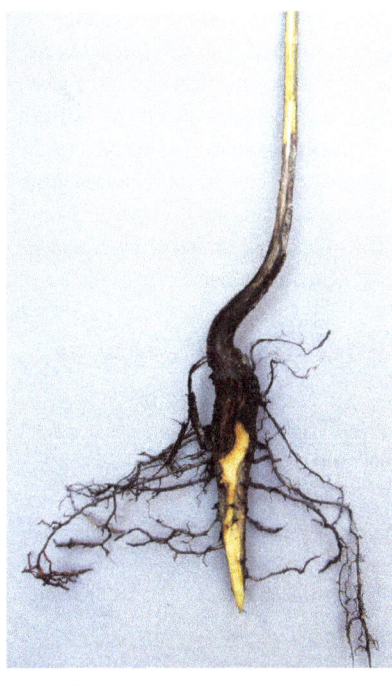

Figure 37.4—*Blackened and water-soaked tissue of black walnut infected by* Phytophthora. Photo by Michelle M. Cram, USDA Forest Service.

Accuracy of these tests is high, but false negatives and positives for *Phytophthora* can occur. To increase test accuracy, take multiple samples and use the diseased tissue along the leading infection edge to avoid secondary organisms that invade dead tissue. Pathologists can provide a very accurate *Phytophthora* species identification with the use of selective culture media; however, laboratory isolation techniques require more time.

Biology

Phytophthora species produce motile spores (zoospores) in sporangia (fig. 37.5) that can swim up to several hours in saturated soil or flowing water before infecting roots or root collars. These motile spores can be produced under wet conditions within 24 hours. Seedlings

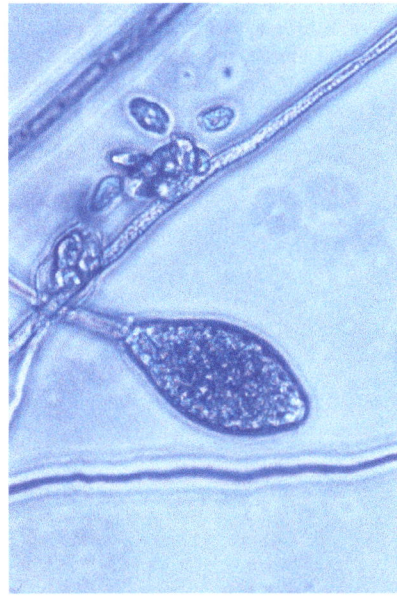

Figure 37.5—*Sporangium and zoospores of* Phytophthora. Photo by Michelle M. Cram, USDA Forest Service.

growing in flooded or saturated soil are further predisposed to infection from damage created by oxygen deficiency. *Phytophthora* species can survive dry conditions and temperature extremes by forming resting structures (chlamydospores and oospores) in damaged plants or soil. These resting structures will remain dormant until activated by high soil moisture.

Control

Prevention

Select nursery fields with well-drained soils. Maintain good drainage by avoiding compaction and promptly fixing leaking irrigation pipes and sprinkler heads. Avoid transplanting seedlings from infested to disease-free nurseries or fields. Do not use diseased trees as mulch and avoid moving equipment from infested to uninfested areas. Use *Phytophthora*-free water sources, such as well water, if possible. Irrigation water from sources that include surface runoff may have *Phytophthora* contamination and require chlorination before use.

Cultural

Reduce potential for saturated soil by improving soil surface and subsurface drainage. Drainage can be improved by building raised nursery beds or installing underground drain tile systems and drainage ditches. Correct compacted soils by deep ripping or chiseling. After seedbeds are sown, wrenching can help reduce compaction. Prevent overwatering by monitoring soil moisture and irrigation output.

Patches or small areas of seedlings with Phytophthora root rot should be lifted separately from the noninfested areas and destroyed. Any equipment used in infested areas must be cleaned of soil

Conifer and Hardwood Diseases

37. Phytophthora Root Rot

and plant debris before being used again. When possible, avoid moving equipment through areas affected by *Phytophthora,* or work in infested areas last and clean the equipment before moving to another field. Losses to Phytophthora root rot during storage and outplanting can be reduced by not lifting seedlings in infested zones and culling seedlings that appear diseased.

In container operations, avoid *Phytophthora* contamination by irrigating with uninfested water and keeping containers and soil off the ground. Use a potting mix that is highly porous and avoid soil compaction in the containers.

Chemical

Fumigation significantly reduces *Phytophthora* populations in the soil, although it is less effective in poorly drained soils where *Phytophthora* is most likely to be a problem. Systemic fungicide use may also be required to control Phytophthora root rot development, especially with susceptible seedlings in nurseries affected by this pathogen in the past. Fungicides are more effective if applied prior to infection and can be used to reduce disease progression in an infested field. They seldom eradicate an established infection, and may make diagnosis more difficult by temporarily limiting symptom expression on already infected seedlings. Rotate between different chemical families to avoid development of resistant strains of *Phytophthora* to systemic fungicides.

Selected References

Bruck, R.I.; Kennerly, C.N. 1983. Effects of metalaxyl on *Phytophthora cinnamomi* root rot of *Abies fraseri*. Plant Disease. 67: 688–690.

Cooley, S.J. 1987. Management of Phytophthora root rot in conifer nurseries of the Pacific Northwest. Tree Planters' Notes. 38: 37–40.

Green, R.J. 1975. Phytophthora root rot of black walnut seedlings. In: Cordell, C.E.; Anderson, R.L.; Hoffard, W.H.; Landis, T.D.; Smith, Jr., R.S.; Toko, H.V., tech. coords. Forest nursery pests. Agriculture Handbook 680. Washington, DC: USDA Forest Service: 99–100.

Green, R.J.; Pratt, R.G. 1970. Root rot of black walnut seedlings caused by *Phytophthora citricola*. Plant Disease Reporter. 54: 583–585.

Hansen, E.M.; Hamm, P.B.; Julis, A.J.; Roth, L.F. 1979. Isolation, incidence, and management of *Phytophthora* in forest tree nurseries in the Pacific Northwest. Plant Disease Reporter. 63(7): 607–611.

Hansen, E.M.; Wilcox, W.F.; Reeser, P.W. 2009. *Phytophthora rosacearum* and *P. sansomeana*, new species segregated from the *Phytophthora megasperma* "complex." Mycologia. 101: 129–135.

Kuhlman, E.G.; Grand, L.F.; Hansen, E.M. 1989. Phytophthora root rot of conifers. In: Cordell, C.E.; Anderson, R.L.; Hoffard, W.H.; Landis, T.D.; Smith, Jr., R.S.; Toko, H.V., tech. coords. Forest nursery pests. Agriculture Handbook 680. Washington, DC: USDA Forest Service: 60–61.

Lane, C.R.; Hobden, E.; Walker, L.; Barton, V.C.; Inman, A.J.; Hughes, K.J.D.; Swan, H.; Colyer, A.; Barker, I. 2007. Evaluation of a rapid diagnostic field test kit for identification of *Phytophthora* species, including *P. ramorum* and *P. kernoviae* at the point of inspection. Plant Pathology. 56: 828–835.

Zentmyer, G.A. 1980. *Phytophthora cinnamomi* and the diseases it causes. Monograph 10. St. Paul, MN: The American Phytopathological Society. 96 p.

Conifer and Hardwood Diseases

38. Pythium Root Rot
Jerry E. Weiland

Hosts

Pythium root rot, caused by various *Pythium* species, affects a wide range of hosts, including most conifer and hardwood seedlings. The most common species identified as causing damage are *P. aphanidermatum*, *P. irregulare*, *P. mamillatum*, *P. splendens*, *P. sylvaticum*, and *P. ultimum*.

Distribution

Pythium species are present in agricultural and nursery soils throughout the United States and Canada.

Damage

Aboveground symptoms include leaf or needle chlorosis, stunting, and seedling death. In some cases, new growth may wilt and form a shepherd's crook before ultimately dying (fig. 38.1). Belowground, roots are stunted with few lateral and feeder roots (fig. 38.2). Damage is often limited to fine roots, although in severe cases necrosis may extend into older root tissues.

Diagnosis

Look for patches of dead or dying seedlings in nursery beds, particularly in low-lying areas where water accumulates (fig. 38.3). Seedlings may pull easily from the ground as the root system deteriorates. The remaining roots are brown or black and often lack healthy, white root tips. Frequently, the cortical tissue is easily slipped off the root, which leaves a white cylinder of xylem (fig. 38.4).

Diagnosis is confirmed through isolation of the pathogen into culture. Alternatively, ELISA (Enzyme-linked immunosorbent assay) test kits, which

Figure 38.1—*Shepherd's crook on seedling killed by* Pythium *species.* Photo by Jerry E. Weiland, USDA Agricultural Research Service.

use specific antibodies to detect *Pythium* species, are available and provide quick results in the field. *Pythium* species, grow readily from freshly killed root tips. Cultures are white and fast-growing with relatively large, coenocytic hyphae. Identification is based on microscopic features or DNA (Deoxyribonucleic acid) analysis. However, fast-growing *Pythium* cultures can mask the presence of slower-growing pathogens such as *Fusarium* or *Phytophthora* species, which also cause root rot. Therefore, the appropriate selection of isolation media is often required for proper diagnosis.

Conifer and Hardwood Diseases

38. Pythium Root Rot

Figure 38.2—*Five-week-old seedlings: Three healthy seedlings on left with a long, central root and lateral roots beginning to form. Three seedlings on right infected with* Pythium ultimum *and showing stunted root and shoot formation.* Photo by Jerry E. Weiland, USDA Agricultural Research Service.

Samples should be collected as soon as possible after symptoms are noticed. Entire plants, including the root system, should be dug, gently shaken to remove excess soil, and then kept cool and moist until the diagnosis can be performed. A collection of seedlings with a range of symptoms is ideal for analysis. Avoid collecting seedlings, however, which have been dead for an extensive period of time because secondary, saprophytic fungi quickly colonize freshly killed plant tissue and can make pathogen isolation difficult.

Biology

Pythium species survive in the soil and crop debris primarily as thick-walled oospores. Root and seed exudates stimulate the oospores to germinate. Hyphae then grow through the soil and directly infect host roots. Fine roots, or roots damaged during planting or weeding, are particularly susceptible. Sporangia are formed where excess moisture is present. Sporangia produce motile spores (zoospores) that swim to the host plant, then attach, encyst, germinate, and infect the root or seed. Excess moisture also allows exudates to travel further in the soil and may stimulate spore production. *Pythium* species are spread through the movement of infested soil or infected plant material and by water splash, runoff, and contaminated irrigation water.

Control

Cultural

Avoid poorly drained soils and low-lying areas that collect water. Do not overwater and install drainage as necessary. Water from recycling ponds should be filtered or treated before reapplication. Avoid excessive fertilization and overcrowding. These conditions promote succulent growth that is susceptible to infection.

Rotation to bare fallow during the summer months can significantly reduce *Pythium* populations. Start with clean stock and avoid replanting newly fumigated fields with heavily infested stock material. In container production systems, start with pathogen-free media or pasteurized soil and use clean, disinfested pots.

Figure 38.3—*Pythium root rot in low-lying area next to leaky irrigation line.* Photo by Jerry E. Weiland, USDA Agricultural Research Service.

Conifer and Hardwood Diseases

38. Pythium Root Rot

Figure 38.4—*Root cortex of a diseased seedling easily slips and reveals a white cylinder of xylem underneath (arrows).* Photo by Jerry E. Weiland, USDA Agricultural Research Service.

Chemical

Management with fungicides requires an accurate diagnosis. *Pythium* species are oomycetes and many fungicides are ineffective against these particular pathogens; therefore, make sure to select fungicides that target oomycetes. Rotate fungicide use according to mode of action to prevent resistant *Pythium* strains from developing. Application is achieved through soil drenches or by incorporating granular or wettable powder formulations directly into the soil. Fungicides are best used as a preventative. After seedlings have become infected, fungicides rarely function effectively as a curative treatment.

Fumigation continues to be the most viable method for large-scale disinfestation of field soils. When applied properly, low permeability plastic films (totally impermeable film, TIF) cause fumigants to be retained in the soil for longer periods of time. This may enable nursery managers to reduce fumigant rates while maintaining the same level of disease control observed under conventional rates of application.

Selected References

Hansen, E.M.; Myrold, D.D.; Hamm, P.B. 1990. Effects of soil fumigation and cover crops on potential pathogens, microbial activity, nitrogen availability, and seedling quality in conifer nurseries. Phytopathology. 80: 698–704.

Jones, R.K.; Benson, D.M. 2003. Diseases of woody ornamentals and trees in nurseries. St. Paul, MN: APS Press. 482 p.

Martin, F.N. 2003. Development of alternative strategies for management of soilborne pathogens currently controlled by methyl bromide. Annual Review of Phytopathology. 41: 325–350.

Martin, F.N.; Loper, J.E. 1999. Soilborne plant diseases caused by *Pythium* spp.: ecology, epidemiology, and prospects for biological control. Critical Reviews in Plant Science. 18: 111–181.

Sinclair, W.A.; Lyon, H.H. 2005. Diseases of trees and shrubs. 2nd ed. Ithaca, NY: Cornell University Press. 660 p.

Conifer and Hardwood Diseases

39. Seed Fungi
Stephen W. Fraedrich and Michelle M. Cram

Hosts

Seedborne fungi can potentially affect seed quality of any forest tree species. Some seedborne fungi are saprophytes and have no affect on seed quality, but others are pathogenic and can cause germination failure or seedling diseases. The best known and most studied seedborne fungi are those that affect conifer species. *Diplodia pinea, Lasiodiplodia theobromae, Sirococcus conigenus,* and *Fusarium* species such as *F. circinatum* and *F. oxysporum* are pathogenic fungi routinely associated with conifer seeds. The dogwood anthracnose pathogen, *Discula destructiva* has been found on dogwood seeds. Other species in genera such as *Pencillium, Aspergillus, Pestalotia, Trichoderma, Mucor,* and *Rhizopus* are associated with various tree seeds, although their role in seed deterioration and quality is often not clear.

Distribution

Seedborne fungi are widely associated with forest tree seeds in all regions of the United States. Specific fungi, however, can be limited in their range because of climatic conditions and host type. For example, *F. circinatum* (pitch canker fungus) has been associated with southern pine and Monterey pine seeds in California maritime areas. In contrast, *S. conigenus* and *C. fulgens* are only associated with conifer seeds in the Northern United States.

Damage

Some seedborne fungal pathogens primarily cause disease in seeds and appear to have little-to-no effects on other developmental stages of trees. These pathogenic fungi infect the internal tissue of seeds, destroying the embryo and endosperm or gametophyte tissue (figs. 39.1 and 39.2). Examples include *L. theobromae,* which destroys slash pine seeds in the Southern United States, and *C. fulgens,* which affects seed quality of conifers in Canada and the Northern United States. In contrast, there are several notable conifer pathogens found on seeds that may have little effect until after seed germination, when they can cause damping-off, root rot, and seedling blight (fig. 39.3). These pathogens frequently reside on or in the seedcoat and initially cause little damage to the seeds. Occasionally, seedborne pathogens will infect the cotyledons of developing seedlings that have the seedcoat still attached (fig. 39.4). Some pathogens of this type can have severe economic and ecological consequences if they are introduced into regions of the world where they are not native. These pathogens include *D. pinea* (syn. *Sphaeropsis sapinea*), *S. conigenus, F. circinatum,* and *F. oxysporum.*

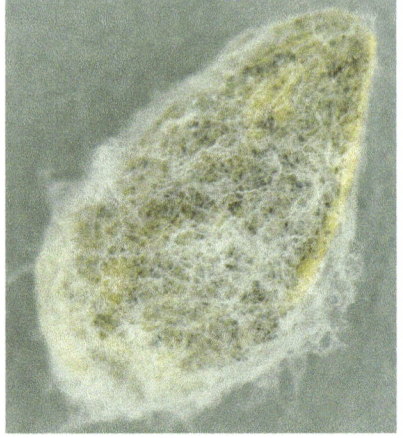

Figure 39.1—*Hyphae of a fungus on the seedcoat of a contaminated pine seed after incubation in the laboratory.* Photo by Stephen W. Fraedrich, USDA Forest Service.

Figure 39.2—*Black seed rot of slash pine caused by* Lasiodiplodia theobromae. Photo by Stephen W. Fraedrich, USDA Forest Service.

Figure 39.3—*Mortality of longleaf pine seedlings caused by the pitch canker fungus* (Fusarium circinatum) *that was associated with contaminated seed sources.* Photo by Stephen W. Fraedrich, USDA Forest Service.

Conifer and Hardwood Diseases

39. Seed Fungi

Figure 39.4—*Primary needles of a pine seedling infected by a seedborne fungus.* Photo by Michelle M. Cram, USDA Forest Service.

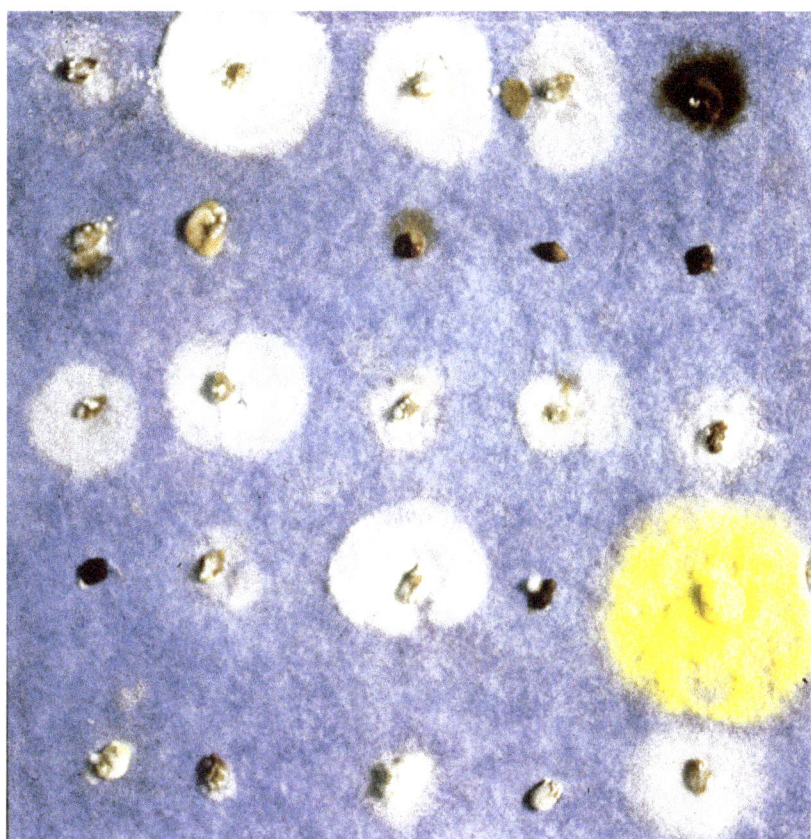

Figure 39.5—Fusarium *species and other fungi growing from seeds placed on a sterile blotter with nutrient growth media.* Photo by Michelle M. Cram, USDA Forest Service.

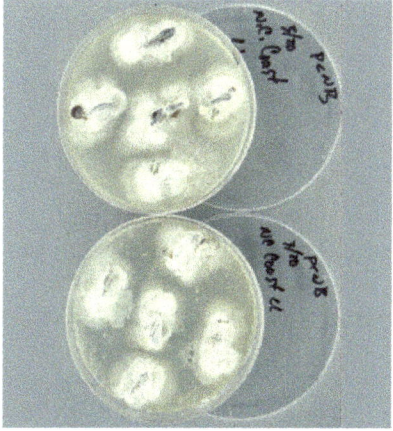

Figure 39.6—Fusarium *species growing from longleaf pine seeds placed on an agar media selective for* Fusarium *species.* Photo by Stephen W. Fraedrich, USDA Forest Service.

Diagnosis

Seed diseases can go unnoticed until an unusually high germination failure or damping-off episode occurs in the nursery. Seedlot germination tests can screen for potential seed disease problems. The incidence of seeds with internal damage caused by pathogenic fungi can be determined for individual seedlots by cutting open a sample of seeds and examining for presence of fungal mycelium. Radiographic techniques can also be used to detect internal seed damage caused by fungal pathogens. Specific fungi associated with seeds can be determined by pathologists who can process samples on artificial media and examine the samples microscopically (figs. 39.5 and 39.6). Although this procedure is widely employed, it is time consuming and may not always detect pathogens at low levels. Efforts are being made to develop molecular tools for detection of pathogens such as *F. circinatum* and *D. pinea*. Molecular techniques could prove to be less time consuming for making diagnoses and may provide more sensitive techniques for screening larger samples of seeds for specific pathogens.

Biology

Seeds may be infected internally, resulting in the destruction of endosperm and embryo or externally on the seedcoat. Fungi exist in seeds as spores and hyphae, and they can survive for long time periods on the seedcoat and in the internal, diseased seed tissue. These features make seeds an ideal means for the long distance transport of pathogens.

Forest Nursery Pests 133

Conifer and Hardwood Diseases

39. Seed Fungi

Seedborne fungi such as *Penicillium, Aspergillus, Mucor,* and *Rhizopus* are often regarded as saprophytes, and their association with seeds is usually an indication of poor seed quality. These fungi often colonize seeds as a result of physical damage or adverse environmental conditions such as high temperatures or moisture conditions that may affect seeds during collection, extraction, and storage.

Other fungi such as *S. conigenus, F. circinatum,* and *D. pinea* are seedborne pathogens capable of causing other diseases such as stem cankers and shoot dieback. *S. conigenus* becomes established in seedlots when older cones are inadvertently included in the cone harvest. The conditions that favor this pathogen's establishment in seeds include high humidity, low light, and cool temperatures ranging from 10 to 20 °C (50 to 68 °F). Seed contamination levels by *Fusarium* species can vary by collection date and by orchard. Fresh wounds are known to provide infection courts for *F. circinatum* and various agents can wound reproductive structures, including insects, high wind, hail, and cone handling practices. Seed contamination by *F. circinatum* can also vary greatly among clones in seed orchards. Cones from clones that are highly susceptible to pitch canker tend to have greater levels of seed contamination. These diseased cones can contaminate the cones and seeds from less susceptible clones when they are included in bulk collections. Not all *Fusarium* species are pathogenic to tree seeds and isolates of some species (for example, *F. oxysporum*) can vary greatly in pathogenicity.

Control

Prevention

Proper cone and seed collection, storage, and extraction techniques are essential to seed disease prevention. Some seedborne diseases can be prevented by not collecting old cones or cones that have been in contact with the ground for extended periods. The potential for seed diseases can also be reduced by collecting mature cones with a specific gravity less than 0.89 and storing them in dry, aerated conditions. Avoid collecting diseased and damaged cones, or sowing highly infested seedlots.

Cultural

Collecting seeds by clone may help to reduce the incidence of seedborne contamination in clones unaffected by the disease. Mechanical seed treatments, such as specific gravity tables and the IDS (Incubation Drying Separation) system, can be used to remove fungus-damaged seeds from seedlots. Avoid damaging seeds during cleaning, dewinging, and treatment. Seed treatments such as water rinses and heat treatments have been shown to control some seedborne pathogens without deleterious effects on seed germination. Although these treatments may reduce the incidence of seedborne fungi, they may not be as effective as some chemical seed treatments.

Chemical

Seed treatments primarily reduce or remove pathogens from the seed coat. Nurseries commonly use labeled fungicides, and in some cases disinfectants, as seed treatments. Seed treatments can have a negative effect on germination and should be applied with caution to avoid unnecessary damage. Disinfectants such as sodium hypochlorite, hydrogen peroxide, and hydrogen dioxide can be used to reduce fungal contamination and improve seed germination. Hydrogen peroxide has long been known to eliminate seedborne mycoflora and stimulate seed germination.

Selected References

Anderson, R.L.; Miller, T. 1989. Seed fungi. In: Cordell, C.E.; Anderson, R.A.; Hoffard, W.H.; Landis, T.D.; Smith, Jr., R.S.; Toko, H.V., tech. coords. Forest nursery pests. Agriculture Handbook 680. Washington, DC: USDA Forest Service: 126–127.

Cram, M.M.; Fraedrich, S.W. 2010. Seed disease and seedborne pathogens of North America. Tree Planters' Notes. 53(2): 35–44.

Sutherland, J.R.; Miller, T.; Quenard, R.S. 1987. Cone and seed disease of North American conifers. North American Forestry Commission. Pub. 1. Victoria, British Columbia, Canada. 77 p.

USDA Forest Service. 2008. The woody plant seed manual. Agriculture Handbook 727. Washington, DC: USDA Forest Service. 1233 p.

Conifer and Hardwood Diseases

40. Sudden Oak Death
Gary Chastagner, Marianne Elliott, and Kathleen McKeever

Hosts

The exotic, federally quarantined plant pathogen *Phytophthora ramorum* causes Sudden Oak Death (fig. 40.1) and ramorum shoot or leaf blight on more than 100 hosts from 36 different families. Hosts include a number of conifers, maples, tanoak, beech, and oak species (table 40.1). Although some hosts, such as oak and tanoak, develop stem cankers, most hosts only develop leaf spots and twig dieback when infected by *P. ramorum*, and are not usually killed by the pathogen. These diseases are commonly referred to as ramorum blight or dieback. In addition to hosts that have been naturally infected in the field, laboratory research indicates that a number of conifers, particularly many true firs and larch, may also be potential pathogen hosts.

Distribution

Naturally infected hosts have been reported in forests and landscapes in California and southwestern Oregon coastal areas. Infected ornamental nursery stock, has been detected throughout the United States and British Columbia, Canada. Despite efforts to eradicate the pathogen from nurseries, the pathogen has spread to waterways and plants outside of infected nurseries at a limited number of sites. The distribution of this disease is expected to increase over time.

Damage

Depending on the host species affected, *P. ramorum* may cause leaf spots, shoot blight, leaf and twig dieback, and outright tree mortality from stem cankers. Besides the direct damage this pathogen may cause, Federal and State regulatory actions associated with the detection of infected host

Figure 40.1—*Tanoaks killed by* Phytophthora ramorum *in a California forest.* Photo by Gary Chastagner, Washington State University.

Table 40.1—*Partial list of forest tree species that are hosts of* Phytophthora ramorum.

Name	Species	Symptom type
Tanoak	*Notholithocarpus densiflorus*	Stem and foliage
Coast live oak	*Quercus agrifolia*	Stem
California black oak	*Quercus kelloggii*	Stem
Shreve's oak	*Quercus parvula* var. *shrevei*	Stem
Canyon live oak	*Quercus chrysolepis*	Stem
Douglas-fir	*Pseudotsuga menziesii*	Foliage, stem, and shoot dieback
Grand fir	*Abies grandis*	Foliage and shoot dieback
White fir	*Abies concolor*	Foliage and shoot dieback
California red fir	*Abies magnifica*	Foliage and shoot dieback
Coast redwood	*Sequoia sempervirens*	Foliage
European yew	*Taxus baccata*	Foliage
Pacific yew	*Taxus brevifolia*	Stem, foliage, and shoot dieback
California nutmeg	*Torreya californica*	Foliage and shoot dieback
Japanese larch	*Larix kaempferi*	Stem, foliage, and shoot dieback
Western hemlock	*Tsuga heterophylla*	Stem, foliage, and shoot dieback
California bay laurel	*Umbellularia californica*	Foliage
Pacific madrone	*Arbutus menziesii*	Foliage
Horse chestnut	*Aesculus hippocastanum*	Stem
Oregon ash	*Fraxinus latifolia*	Foliage
Bigleaf maple	*Acer macrophyllum*	Foliage

Conifer and Hardwood Diseases

40. Sudden Oak Death

material have the potential for significant economic losses to individual nurseries and the forest nursery industry.

Diagnosis

Symptoms caused by *P. ramorum* on conifer hosts look much like injury from a late frost or Botrytis tip blight. Initial infections occur on new growth in spring, and during bud break and shoot elongation periods. After the initial infection, the pathogen causes the emerging shoot to wilt (fig. 40.2). The pathogen may continue to spread down the shoots, resulting in dieback. Dieback extent varies by host, when the infection occurs, and environmental conditions. When shoots are infected just after bud break, the pathogen commonly grows down the shoot into the previous year's growth. Often the needles on the previous year's growth are shed as the pathogen grows down the shoot, resulting in dead twigs with tufts of dead terminal needles (fig. 40.3). If infection occurs later during shoot elongation, the pathogen may only cause a dieback of the newly developing shoot. Repeated infection may kill seedlings and saplings or greatly alter young tree growth and form. On conifer nursery stock, where shoot elongation often occurs throughout the growing season, it is likely that *P. ramorum* infections are not limited to the spring. Symptoms on hosts other than conifers can vary greatly, from leaf spots to mature tree death.

Phytophthora species produce no easily recognizable fruiting bodies or spores on infected tissue, and diagnosis can be confirmed only through isolating the organism into pure culture. Because *P. ramorum* causes symptoms similar to those caused by other plant pathogens and abiotic conditions, samples must be sent to a lab for diagnosis. Selective media are often used in attempts to isolate the pathogen from infected tissues. *P. ramorum* produces copious amounts of chlamydospores in culture (fig. 40.4). *Phytophthora* species identification is usually left to specialists. Polymerase chain reaction (PCR) molecular tests are often used to confirm *P. ramorum* presence in samples.

Biology

During wet periods, *P. ramorum* produces copious amounts of sporangia and zoospores from spots on infected leaves of epidemiologically important (spore-producing) hosts such as California bay laurel (fig. 40.5). The spores are then carried by wind-driven rain to the foliage and bark of other susceptible hosts. Research on Christmas tree varieties in California shows that fairly high inoculum (spores) levels are necessary for Douglas-fir infection, and that little infection risk exists for trees located 5 to 8 m (16 to 26 ft) away from infected, epidemiologically important hosts. Recent observation and research in the United Kingdom indicates that infected Japanese larch support high foliar sporulation levels, unlike Douglas-fir and grand fir, which are most likely epidemiologically unimportant because inoculum is not produced on infected tissues. In addition to its natural dispersal, *P. ramorum* can also be spread via contaminated irrigation water and the movement of contaminated soil and infected plants.

This pathogen grows between 2 and 28 °C (35 and 82 °F), with optimum growth at 20 °C (68 °F). Hyphae can survive short time periods at temperatures as low as -5 °C (23 °F) and as high as 30 °C (86 °F). Chlamydospores embedded in host tissue can survive temperature extremes from -10 to 35 °C (14 to 95 °F) for up to 1 week.

Figure 40.2—*Tip dieback of Douglas-fir caused by* Phytophthora ramorum *in a California Christmas tree plantation.* Photo by Gary Chastagner, Washington State University.

Figure 40.3—*Tip dieback of grand fir caused by* Phytophthora ramorum *in a California Christmas tree plantation. Photo was taken in May shortly after infection following spring rainstorms.* Photo by Gary Chastagner, Washington State University.

Conifer and Hardwood Diseases

40. Sudden Oak Death

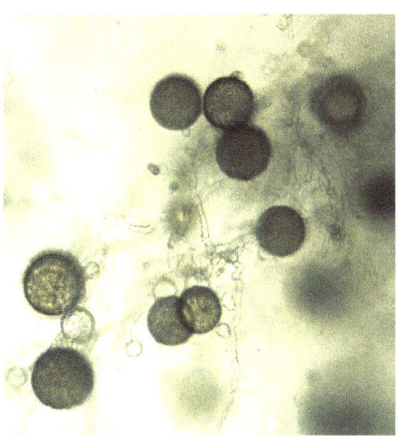

Figure 40.4—*Chlamydospores and hyphae of* Phytophthora ramorum *as seen in culture.* Photo by Marianne Elliott, Washington State University.

Figure 40.5—*Tips of California bay laurel leaves infected with* Phytophthora ramorum. *This host is responsible for much of the disease spread in California forests.* Photo by Kathy Riley, Washington State University.

In water, zoospore release from sporangia is stimulated by exposure to low temperatures around 4 °C (39 °F), and direct sporangia germination occurs at higher temperatures.

Control

Prevention

Although the risk of *P. ramorum* developing on conifer nursery stock is unclear, excluding the pathogen from the nursery is the single most effective way to reduce risks. Do not transplant seedlings from infested nurseries into disease-free ones. In infested areas, a seedling in close proximity to a spore-producing host has an increased *P. ramorum* infection risk. In these areas, nursery perimeters should be inspected for the presence of inoculum-producing hosts that could potentially serve as an inoculum source. Inform crews about *P. ramorum* implications on nursery stock and sanitation practices that reduce pathogen transfer when working in infested areas. Avoid equipment and crew movement between infested and noninfested areas. Scrape, brush, and hose off accumulated soil and mud from clothing, gloves, boots, and shoes and sanitize these items with a disinfectant.

Cultural

Remove potential landscape hosts that are located within 10 m (33 ft) of the nursery site. Don't irrigate crops with water from streams or ponds potentially contaminated with the pathogen unless the water has been treated. Several methods for treating *Phytophthora*-infested water can be employed, which include adding chemicals such as chlorine to the water and filtrating the water.

Chemical

Several fungicides are registered for *P. ramorum* control in nurseries. Soil fumigation is an effective treatment for soil infested with chlamydospores.

Selected References

California Oak Mortality Task Force. http://www.suddenoakdeath.org (accessed October 2010).

Goheen, E.M.; Hansen, E.; Kanaskie, A.; Osterbauer, N.; Parke, J.; Pscheidt, J.; Chastagner, G.A. 2006. Sudden Oak Death and *Phytophthora ramorum*: a guide for forest managers, Christmas tree growers, and forest-tree nursery operators in Oregon and Washington. Oregon State University Extension Service Pamphlet, EM 8877. 16 p.

Ufer, T.; Posner, M.; Wessels, H.P.; Wagner, S.; Kaminski, K.; Brand, T.; Werres, S. 2008. Four years of experience with filtration systems in commercial nurseries for eliminating *Phytophthora* species from recirculation water. In: Frankel, S.J.; Kliejunas, J.T.; Palmieri, K.M., tech. coords. Proceedings of the Sudden Oak Death Third Science Symposium. PSW-GTR-214. Albany, CA: USDA Forest Service, Pacific Southwest Research Station. 491 p.

USDA Animal & Plant Health Inspection Service (APHIS). http://www.aphis.used.gov/ppq/ispm/sod.

Washington State University. http://www.puyallup.wsu.edu/ppo/gallery/conifers/album/index.html (accessed October 2010).

Werres, S.; Marwitz, R.; Man In't Veld, W.A.; De Cock, A.W.A.M.; Bonants, P.J.M.; DeWeerdt, M.; Themann, K.; Ilieva, E.; Baayen, R.P. 2001. *Phytophthora ramorum* sp. nov., a new pathogen on *Rhododendron* and *Viburnum*. Mycological Research. 105(10): 1155–1165.

Yakabe, L.E.; MacDonald, J.D. 2010. Soil treatments for the potential elimination of *Phytophthora ramorum* in ornamental nursery beds. Plant Disease. 94: 320–324.

Conifer and Hardwood Diseases

41. Yellows or Chlorosis
Tom Starkey

Revised from chapter by Samuel J. Rowan, 1989.

Hosts

All green plants are susceptible to foliage yellowing or chlorosis.

Distribution

Seedling foliage chlorosis occurs throughout the temperate regions of the world.

Damage

The term yellows, or chlorosis, describes a generalized yellowing or bleaching of foliage due to a lack of chlorophyll. The destruction or reduced synthesis of chlorophyll may be caused by any number of biotic or abiotic agents. Chlorosis, for a brief period, may cause slight growth reductions. If chlorosis persists over an extended period of time, however, plant mortality may occur.

Diagnosis

Chlorosis of seedlings may be caused by one or a combination of biotic and abiotic factors, which requires a diagnostic procedure that involves a process of elimination. Look for standing water, insects, heat or cold injury, and fungal disease symptoms. Foliar and soil analysis will aid in pinpointing any nutritional deficiencies, excesses, and problems with soil pH or nematodes.

Biology

A number of factors cause the symptom called yellows or chlorosis. The following list includes some of the factors.

Most Common

1. Nutrient imbalance is a deficiency (or excess) of elements essential to plants such as iron, nitrogen, phosphorus, potassium, and calcium, along with minor elements such as magnesium, manganese, zinc, boron, copper, molybdenum, and sulfur. Nutrient deficiency is generally tied to either high or low soil pH (fig. 41.1). Probably the most common form of chlorosis is called iron chlorosis. Iron chlorosis results when seedlings are unable to obtain iron needed for the production of chlorophyll. In conifers, chlorosis is observed as an overall yellowing of needles. In mild cases, the younger needles are most affected. In more severe cases, the whole plant may become chlorotic. In hardwoods, the primary symptom is interveinal chlorosis (fig. 41.2)—observable as brown spots that develop between the main leaf veins. Iron chlorosis generally begins on the newer growth at the branch tips. If not corrected, the leaves may curl, dry up, and drop off.

2. Adverse environmental conditions or factors that can cause chlorosis include extreme temperatures and either excess or inadequate soil drainage. Cold temperatures, for example, are often associated with pigment synthesis other than chlorophyll, leading to red, purple, or yellow seedlings. Seedling color may return to a normal green when conditions improve as long as these are short-term environmental stresses.

3. Parasitic fungi, bacteria, and nematodes cause root, stem, or foliage disease. These pathogens are commonly associated with chlorotic seedlings. In most cases, roots are destroyed, thereby limiting the seedlings' ability to take up essential elements (fig. 41.3).

Figure 41.1—*Chlorosis caused by nutrient deficiency in a southern pine nursery.* Photo by David B. South, Auburn University.

Conifer and Hardwood Diseases

41. Yellows, or Chlorosis

Figure 41.2—*Interveinal chlorosis of red maple caused by iron deficiency.* Photo by John Ruter, University of Georgia, at http://www.bugwood.org.

Least Common

1. Toxic concentrations of herbicides, fungicides, nematicides, insecticides, and other compounds used in nurseries.
2. Feeding by insects and mites, such as red spider mites on coniferous seedlings.
3. Certain viruses and mycoplasmas that primarily affect hardwood tree species.
4. Genetic abnormalities, which result in the loss of ability to synthesize chlorophyll. Albinism is the most common genetic abnormality associated with chlorotic forest tree seedlings. These seedlings generally do not survive much past the germination phase.
5. Soils high in soluble salts.

Control

Cultural

Adjust soil pH to between 5.0 and 6.0 by either adding lime to raise the pH or ammonium sulfate (or other sulfur compounds) to lower the soil pH. Adjust nutrient deficiencies by applying the required mineral element(s) to the foliage or soil. Reduce the effects of air and soil temperatures by mulching, shading, and irrigating seedbeds. Ensure the nursery beds drain properly to eliminate excess soil water.

Chemical

Root diseases and nematodes can be controlled by soil fumigation prior to sowing. Use deep shank injection of soil fumigants to provide better control of persistent nematode problems. Foliage and stem pathogens can be controlled with fungicidal foliar sprays. Insects and mites can be controlled with insecticides.

Selected References

Filer, Jr., T.H.; Cordell, C.E. 1983. Nursery diseases of southern hardwoods. USDA Forest Service Forest Insect & Disease Leaflet 137. Washington, DC. 6 p.

Hacskaylo, J.; Finn, R.E.; Vimmerstedt, J.P. 1969. Deficiency symptoms of some forest trees. Res. Bull. 1015. Wooster, OH: Ohio Agricultural Research and Development. 69 p.

Hodges, Jr., C.S.; Ruehle, J.L. 1979. Nursery diseases of southern pines. USDA Forest Service Forest Pest Leaflet 32. Washington, DC. 8 p.

Landis, T.D. 1997. Micronutrients—Iron. Forestry Nursery Notes. July: 12–16.

Mengel, K.; Kirkby, E.A. 1987. Principles of plant nutrition, 4th ed. International Potash Institute, Berne, Switzerland. 745 p.

Nakos, G. 1979. Lime-induced chlorosis in *Pinus radiata*. Plant and Soil. 52: 527–536.

Rowan, S.J. 1989. Yellows, or chlorosis. In: Cordell, C.E.; Anderson, R.L.; Hoffard, W.H.; Landis, T.D.; Smith, Jr., R.S.; Toko, H.V., tech. coords. Forest nursery pests. Agriculture Handbook 680. Washington, DC: USDA Forest Service: 130–131.

Van Dijk, H.F.G.; Bienfait, H.F. 1993. Iron-deficiency chlorosis in Scots pine growing on acid soils. Plant Soil. 153: 255–263.

Figure 41.3—*Nematode-induced chlorosis in second year field (left side of photo) adjacent to first year methyl bromide fumigated field (right side of photo).* Photo by Tom Starkey, Auburn University.

140 Forest Nursery Pests

Conifer and Hardwood Insects

Conifer and Hardwood Insects

42. Aphids
Art Antonelli

Revised from chapter by Michael D. Connor and William M. Hoffard, 1989.

Hosts

Seedlings of all tree species are subject to attack by one or more species of aphids. Species of the genus *Cinara*, which attack conifers, are probably the most important of the seedling-infesting aphids.

Distribution

Aphids occur throughout North America. There are approximately 1,380 species in 277 genera.

Damage

Aphids feed on plant sap, causing a general weakening in the area of feeding. Heavy feeding may result in branch dieback or the death of seedlings. Aphids may also spread pathogenic viruses during feeding. When this occurs, damage to plants can be substantial.

Diagnosis

Aphids attack all types of seedling tissues (roots, stems, and foliage). Look for symptoms such as yellowing of foliage, curling of leaves, branch deformities, and galls. Aphids excrete a sugary substance called honeydew. A fungus, called sooty mold, often grows on the honeydew and makes the foliage appear black. Ants are often found together with aphids; they protect the aphids and feed on the honeydew. The combination of ants and black foliage is a good indication that aphids may be a problem. Aphids are soft-bodied, globe- or pear-shaped insects, about 0.8 to 6.4 mm long (fig. 42.1). They may be white, pink, yellow, green, blue, brown, gray, or black. The adults may be winged or wingless. When wings are present, two pairs fold in a roof-like fashion over the body (fig. 42.2) Aphids have three pairs of thoracic legs. The mouth is a piercing-sucking beak, which appears to arise between the front legs. Most species have a pair of tube-like structures located on the back of the abdomen. Aphids are usually exposed while feeding, but some cover themselves with a waxy material (fig. 42.3), and others feed inside galls.

Biology

Since there are so many species of aphids, only a general description can be given of their life cycles. Winged females lay eggs. The eggs hatch, and nymphs mature into wingless adults. These wingless

Figure 42.2—*Winged aphid with offspring.* Photo by Ken Gray. Image courtesy of Oregon State University.

Figure 42.1—*Cinara species aphids feeding on stem of conifer seedling.* Photo courtesy of Washington State University.

Figure 42.3—*Root aphids on conifer roots. Note white, waxy bodies.* Photo by Art Antonelli, Washington State University.

forms reproduce without mating, and the females usually bear live young. After several generations, winged females are produced. These winged females migrate to a new host, and wingless females are again produced. After several more generations, males and females are produced. They mate, the females lay eggs, and the cycle is repeated. Some species of aphids have a more complicated life cycle and migrate between a primary and secondary host. Environmental conditions strongly affect aphid populations. Fungus-caused diseases; parasitic wasps; and predators like lady beetles, lacewings, nabid bugs, and syrphid fly larvae provide some natural control. When environmental conditions are favorable for the aphid population, however, biological agents cannot keep populations in check. This is especially true in nurseries, where susceptible host plants are concentrated in a small area.

Control

Cultural

Maintain seedling vigor through proper fertilization and irrigation to increase tolerance to aphids.

Chemical

Use registered insecticides to effectively control aphids. Apply sprays to the point of runoff on all foliage and stem surfaces. Read and follow label directions carefully to prevent seedling injury.

Selected References

Anderson, R.F. 1960. Forest and shade tree entomology. New York: John Wiley & Sons: 347–351.

Connor, M.D.; Hoffard, W.H. 1989. Aphids. In: Cordell, C.E.; Anderson, R.L.; Hoffard, W.H.; Landis, T.D.; Smith, Jr., R.S.; Toko, H.V., tech. coords. Forest nursery pests. Agriculture Handbook 680. Washington, DC: USDA Forest Service: 134–135.

Cranshaw, W. 2004. Garden insects of North America. Princeton, NJ: Princeton University Press. 656 p.

Furniss, R.L.; Carolin, V.M. 1977. Western forest insects. USDA Forest Service Misc. Pub. No. 1339. Washington, DC: USDA Forest Service. 654 p.

Hollingsworth, C.S.; Antonelli, A.; Hirnyck, R., eds. 2009. Pacific Northwest insect management handbook. Corvallis, OR: Oregon State University Press. 698 p.

Johnson, N.E. 1965. Reduced growth associated with infestations of Douglas-fir seedlings by *Cinara* species (Homoptera: Aphidae). Canadian Entomologist. 97(2): 113–119.

Johnson, W.T.; Lyon, H.H. 1991. Insects that feed on trees and shrubs, 2nd ed. rev. Ithaca, NY: Cornell University Press: 68–69, 248–271.

Pike, K.S.; Boydston, L.L.; Allison, D.W. 2003. Aphids of western North America with keys to subfamilies and genera for female alatae. Washington State University Extension Miscellaneous 0523. Pullman, WA: Washington State University Extension. 282 p.

Smith, C.F.; Parron, C.S. 1978. An annotated list of aphididae (Homoptera) of North America. Tech. Bull. 255. Raleigh, NC: North Carolina Agricultural Experiment Station. 428 p.

Conifer and Hardwood Insects

43. Cutworms
Michael E. Ostry and Jennifer Juzwik

Hosts and Pest Description

Cutworms (family Noctuidae) feed on the foliage of a wide variety of weeds and agricultural crops. In bareroot tree nurseries, cutworms feed on young conifer and hardwood seedlings. A number of different cutworm species occur in forest nurseries, and the species may differ by geographical region. The adult is a moth (order Lepidoptera) whose head has an "owl-like" appearance. Larvae coloration and size varies among species.

Distribution

Several cutworm species are found throughout the United States.

Damage

Large cutworm populations, such as the dingy cutworm (*Feltia ducens*) (fig. 43.1), can quickly destroy conifer seedlings by feeding on recently emerged seedling cotyledons and needles (fig. 43.2) in tree nurseries. Cutworms can also feed on older seedling foliage, but damage is usually insignificant. Young hardwood seedlings can be cut at or slightly above or below the groundline. Climbing cutworm species can feed on young leaves. Populations fluctuate widely from year to year and since high populations build up only during favorable environmental conditions, cutworms are generally only of periodic importance.

Figure 43.1—*Dingy cutworm larva (left) and pupa.* Photo from USDA Forest Service Archives.

Figure 43.2—*Cutworm damage on young conifer seedling. Note clipped needles.* Photo from USDA Forest Service Archives.

Conifer and Hardwood Insects

43. Cutworms

Diagnosis

Frequently, cutworm damage is noted before observing the insect itself because they feed at night and remain underground during the day. On conifer seedlings look for cut or chewed primary needles. Old chewing damage may become sunken or depressed (fig. 43.3). This damage type may be confused with damping-off disease. Occasionally, both conifer and hardwood seedlings are clipped at the groundline, and tops are left lying on the soil surface. Cutworm larvae are stout, hairless, and dull gray in color, ranging from 2 to 5 cm (0.8 to 2.0 in) at maturity. When disturbed, cutworms drop to the ground and characteristically assume a curled position. The adult moths are hairy, with markings on their forewings but with rather nondescript hind-wings. It is often difficult to identify the many different cutworm species.

Biology

Most cutworm species have similar life cycles. Depending on the geographic location and cutworm species, there may be multiple generations per year. All cutworm species become active in early spring, during the larval stage. They feed on the newly emerging seedlings and rapidly progress through as many as seven instars. Moths emerge in late summer and early fall. Females deposit eggs on plant foliage such as weeds in and around the nursery or in the soil, where eggs or larvae overwinter.

Control

Cultural

Good weed control in and around nursery beds will eliminate breeding sites for cutworms. Emerging seedlings in nursery beds should be examined weekly for feeding evidence.

Chemical

Areas of nursery beds diagnosed with cutworm feeding can be treated with approved insecticides at the first sign of damage. Cutworm species identification may be necessary to determine the appropriate insecticides available for use. Biological insecticides are available for some species.

Selected References

Palmer, M.A.; Hoffard, W.H. 1989. Cutworms. In: Cordell, C.E.; Anderson, R.L.; Hoffard, W.H.; Landis, T.D.; Smith, Jr., R.S.; Toko, H.V., tech. coords. Forest nursery pests. Agriculture Handbook 680. Washington, DC: USDA Forest Service: 136–137.

Palmer, M.A.; Nicholls, T.H. 1981. How to identify and control cutworm damage on conifer seedlings. HT-51. St. Paul, MN: USDA Forest Service, North Central Forest Experiment Station. 6 p.

USDA Forest Service. 1985. Insects of eastern forests. Misc. Pub. No. 1426. Washington, DC: USDA Forest Service. 608 p.

Figure 43.3—*Cutworm damage. Note depressed area on stem.* Photo from USDA Forest Service Archives.

Conifer and Hardwood Insects

44. Field and Short-Tailed Crickets
Coleman Doggett

Hosts

Short-tailed crickets are in the genus *Anurogryllus* and field crickets are in the genus *Gryllus*. Because these two species have similar physical appearance, they are lumped together in this chapter. Short-tailed crickets have been reported feeding on newly germinated slash pine seedlings and field crickets have been observed feeding on eastern redcedar seedlings in the Southeastern United States.

Distribution

The short-tailed cricket is represented in the United States by the single species *A. arboreus*. It is found from New Jersey to Florida and west into Texas and Oklahoma.

A number of field cricket species in the genus *Gryllus* are found in the United States, some distinguished only by their chirping sounds. They have similar eastern and southern distribution as the short-tailed cricket, but their range extends farther north into Massachusetts and as far west as Kansas.

Damage

Short-tailed crickets live in underground burrows and emerge only to feed and find mates. If pine seed or newly germinated seedlings are available, the cricket eats the seed, severs the seedling at the groundline, and drags it into the burrows to consume.

Adult field crickets spend their lives above ground. When their populations reach high proportions in nursery beds and redcedar seedlings are available, the crickets feed on seedling stems. They debark the stems to a height of about 2.5 cm (1 inch), which results in seedling death.

Diagnosis

Adult field crickets' color may be black, brown, or reddish, and about 1.9 cm (0.75 in) long. Females are equipped with spike-like ovipositors at the end of their abdomens, which are almost as long as their bodies (fig. 44.1). Males can be identified by the lack of ovipositors. When high field cricket populations occur in conjunction with redcedar seedlings, look for dying seedlings and check seedlings for debarking at, or slightly above, the groundline.

Adult short-tailed crickets are very similar to field crickets except that females lack a long ovipositor (fig. 44.2). Short-tailed crickets are light brown in color. The crickets leave small soil pellets or mounds similar to crawfish activity. When observed in or near nursery beds, closely examine the beds near the soil pellets for signs of severed seedlings. The tops of the severed seedlings are normally dragged into burrows to be eaten, and the only evidence of injury is the severed seedling stem near the groundline.

Figure 44.1—*Field cricket.* Photo by Richard Grantham, Oklahoma State University.

Conifer and Hardwood Insects

44. Field and Short-Tailed Crickets

Biology

Female short-tailed crickets lay eggs in their burrows in the late spring. The eggs hatch and the resulting nymphs remain in the mother's burrows for about 1 month, after which they disperse for a few yards and dig their own burrows. At first the burrows are small and near the surface, but as the cricket matures, the burrow becomes larger and deeper, eventually reaching depths of 30 to 50 cm (12 to 20 in). Only one cricket is found per burrow, except when newly hatched nymphs are present. There is only one generation per year.

Female field crickets deposit eggs in the soil using their long ovipositors. When the eggs hatch, the nymphs burrow to the surface and emerge. The nymphs are similar to the adults, except they are smaller, wingless, and not sexually mature. Within 2 to 3 months, the nymph molts 8 to 10 times before becoming an adult. Some species overwinter as eggs and others as adults. Populations are very cyclic, low in some years and quite high in others.

All cricket species have many predators including birds and many animal species.

Control

Chemical control has been very effective in halting damage with early detection. The younger the insects, the more effective chemical control will be.

Selected References

Arnold, D.; Reback, E.; Royer, T.; Mulder, P.; Kard, B. 2008. Major horticultural and household insects of Oklahoma. E-918. Oklahoma City: Oklahoma Cooperative Extension Service. 176 p.

Doggett, C.A.; Hawley, V.; Norris, W. 1980. Cricket damage to red cedar seedlings. Tree Planters' Notes. 31(3): 18.

USDA Forest Service. 1985. Insects of eastern forests. Misc. Pub. No. 1426. Washington, DC: USDA Forest Service. 608 p.

Figure 44.2—*Short-tailed cricket.* Photo by Richard Grantham, Oklahoma State University.

Conifer and Hardwood Insects

45. Fungus Gnats
Art Antonelli

Hosts

There are two families of fungus gnats: the Mycetophilidae and the Sciaridae, or dark-winged fungus gnats. The mycetophilids are associated with fungi as larvae. The sciarids are general feeders, feeding mostly on fungi, but a few species attack a wide range of ornamental and vegetable plants. They can be a real problem on greenhouse plants.

Distribution

Fungus gnats are ubiquitous and occur wherever conditions are ambient for development. Any nursery or greenhouse where damp organic soils are used and overwatering is a common practice is fair game for these insects.

Damage

The maggot, or larva, is the damaging stage. However, the adult flies are usually noticed before larval damage to the plant is apparent. When maggots become numerous, they strip plant roots (fig. 45.1).

Diagnosis

Feeding by fungus gnat larvae results in loss of plant vigor and yellowing and wilting of the leaves.

Adult fungus gnats (fig. 45.2) are slender and about 2.5 mm long. Their coloring is typically dull, most being yellowish, brown, or nearly black. In general appearance the flies, especially the mycetophilids, are not unlike small mosquitoes. They have long legs and antennae. They are weak fliers but can run quite rapidly across the soil surface. The mature larvae or maggots are about 5.5 mm long with shiny black head capsules and whitish, somewhat transparent legless bodies (fig. 45.3).

Biology

During the female's 1-week lifetime, she lays 100 or more eggs. The shiny white oval eggs are semitransparent and barely visible to the naked eye. They are laid either singly or in stringed groups of 10 or more on the soil surface, usually near host plants. They hatch in 4 to 6 days. The maggot reaches maturity in about 2 weeks, when it ceases feeding, spins a cocoon, and sheds its skin. After about 1 week, the maggot transforms into a pupa. At the end of the pupal period, the adult fly emerges from the soil and starts the cycle over again. Many overlapping generations are born each year.

Control

Cultural

Fungus gnats live in moist, shady environments with decaying organic material, so cultural practices such as the elimination of dead leaves or other decaying organic material and avoidance of excessive watering will greatly reduce their numbers. The use of a water meter to determine water needs will help in avoiding overwatering.

Figure 45.1—*Damage to fine roots due to feeding by fungus gnat larvae.* Photo by Thomas D. Landis, USDA Forest Service.

Figure 45.2—*Adult fungus gnat.* Photo by Ken Gray. Image courtesy of Oregon State University.

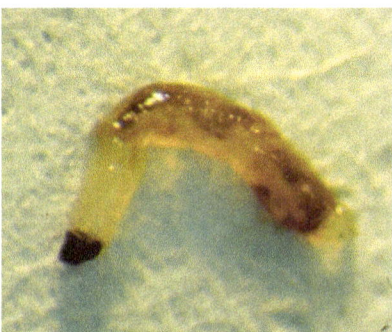

Figure 45.3—*Larva of fungus gnat.* Photo by Thomas D. Landis, USDA Forest Service.

Wherever organic material and moisture accumulates, there is potential for fungus gnats to breed. This condition is particularly true of water drains where such debris can accumulate and provide a breeding site for these flies. A regular (at least once a month) cleaning with a gallon of near boiling water poured down the drain, followed by a cup of bleach diluted with water in a 1:5 ratio, should render this hard-to-reach environment fly and maggot free for 2 weeks in most cases.

Chemical

Control of fungus gnats in commercial plant growing operations, such as greenhouses, can be achieved by using registered formulations of synthetic or biological insecticides.

Monitoring

Successful fungus gnat control depends on a systematic monitoring program for detection of adults. Early detection will result in quicker suppression. For best results, place one yellow sticky trap for every 46 to 93 m^2 (500 to 1,000 ft^2). Check the traps each week, and replace after they become covered with insects.

Selected References

Cole, F.R. 1969. The flies of western North America. Berkeley, CA: University of California Press. 693 p.

Cranshaw, W. 2004. Garden insects of North America. Princeton, NJ: Princeton University Press. 656 p.

Ebeling, W. 1975. Urban entomology. Riverside, CA: University of California—Division of Agricultural Sciences. 695 p.

Triplehorn, C.A.; Johnson, N.F. 2005. Borror and Delong's introduction to the study of insects. 7th ed. Pacific Grove, CA: Thomson-Brooks/Cole. 864 p.

Conifer and Hardwood Insects

46. Lesser Cornstalk Borer
Wayne N. Dixon and Albert E. Mayfield, III

Hosts

The lesser cornstalk borer (*Elasmopalpus lignosellus*) affects seedlings of Arizona cypress, bald cypress, black locust, dogwood, black tupelo, loblolly pine, redcedar, sand pine, slash pine, and sycamore. Agricultural host plants (more than 60 species) include beans, corn, millet, peas, sorghums, and soybeans.

Distribution

The insect is found from Maine to southern California and southward to Mexico, but damage is most severe in nurseries of the Southern United States.

Damage

Complete girdling results in seedling death. Partially girdled seedlings usually recover. Mortality in Arizona cypress may be increased by infection of wounded seedlings by the fungus *Dothiorella* species.

Diagnosis

Look for wounds caused by larval feeding below to just above groundline (fig. 46.1). Bark may be completely or partially removed for up to several centimeters (fig. 46.2). Girdled seedlings remaining alive may have a gall-like swelling on the stem just above the girdle. Partial girdles on the stem are usually closed by callus formation. Severely damaged seedlings die and may remain upright or fall over (fig. 46.3). The slender larvae of the lesser cornstalk borer are about 2.0 cm long when mature. They are pale green and

Figure 46.1—*Feeding wound made by larvae of the lesser cornstalk borer.* Photo by Florida Department of Agriculture and Consumer Services, Division of Forestry.

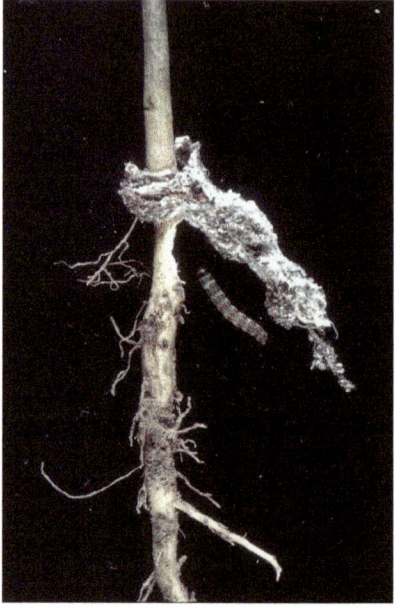

Figure 46.2—*Seedling girdled and debarked by lesser cornstalk borer larvae belowground, showing larva and silken tunnels attached to the stem.* Photo by James D. Solomon, USDA Forest Service, at http://www.bugwood.org.

have brown banding or stripes (fig. 46.4). Silk tunnels, which protect inactive or disturbed larvae, can sometimes be seen radiating from feeding sites (figs. 46.2 and 46.5).

Larvae wriggle furiously when captured, but are difficult to find. Moths may be more readily observed than larvae and are often seen in short and erratic flight patterns just above the seedling tops. They are light- to dark-brownish gray and have a wingspan of approximately 1.6 to 2.4 cm, (figs. 46.6 and 46.7). At rest, female moths are often charcoal-colored and male moths are often tan-colored with charcoal markings.

Conifer and Hardwood Insects

46. Lesser Cornstalk Borer

Figure 46.3—*Fallen bald cypress seedling girdled at the groundline by lesser cornstalk borer.* Photo by Florida Department of Agriculture and Consumer Services, Division of Forestry.

Figure 46.4—*Larva of the lesser cornstalk borer.* Photo by Florida Department of Agriculture and Consumer Services, Division of Forestry.

Biology

The lesser cornstalk borer has two to four generations per year. By late summer, most life stages can be found. After emerging from the soil in late spring, moths mate, and female moths deposit eggs singly in the soil at the bases of host plants or on their stems and lower leaves. Each female lays approximately 125 eggs. Eggs hatch within 1 week, and larvae mine the lowermost branches or begin semisubterranean feeding on stems and roots. Larvae feed from 2 to 3 weeks. Pupation occurs in silk tunnels or soil litter and takes 2 to 3 weeks. Then new adults emerge, mate, and live for about 10 days. Larvae or pupae overwinter in the soil or soil litter.

Figure 46.5—*Silk tunnels formed by larvae of the lesser cornstalk borer, attached to a soybean plant.* Photo by James Castner, Entomology and Nematology Department, University of Florida.

Conifer and Hardwood Insects

46. Lesser Cornstalk Borer

Figure 46.6—*Adult moth of the lesser cornstalk borer, wings folded.* Photo by Gretchen L. Grammer, Grand Bay National Estuarine Research Reserve.

Figure 46.7—*Adult moth of the lesser cornstalk borer, wings extended.* Photo by James T. Vargo.

Control

Cultural

Certain cover crops (for example, soybeans, pearl millet, peanuts, sorghum, peas, and certain grasses), sandy soils, and droughty weather encourage infestations in forest nurseries. Practice general sanitation measures. Fall or winter cleanup of plant residue, early cover crop disking, and unsusceptible cover crop rotation may help reduce the incidence of lesser cornstalk borer.

Chemical

To prevent lesser cornstalk borer incedence, granular insecticides can be applied to the soil when the cover crop is planted. The insecticide used will depend on the cover crop. A remedial, supplementary treatment may also be required. A liquid insecticide formulation may be applied as a soil drench at the first sign of seedling damage. Due to the protective silken tunnels, additional applications may be needed to ensure the larvae are adequately exposed to the insecticide.

Selected References

Davis, T.C.; Goggans, J.F.; Meier, R.J. 1974. Pest control problems encountered in seedling production of Arizona cypress in Alabama. Tree Planters' Notes. 25(2): 7–9.

Dixon, W.N. 1989. Lesser cornstalk borer. In: Cordell, C.E.; Anderson, R.L.; Hoffard, W.H.; Landis, T.D.; Smith, Jr., R.S.; Toko, H.V., tech. coords. Forest nursery pests. Agriculture Handbook 680. Washington, DC: USDA Forest Service: 138–139.

Dixon, W.N.; Barnard, E.L.; Fatzinger, C.W.; Miller, T. 1991. Insect and disease management. In Duryea, M.L.; Dougherty, P.M., eds. Forest regeneration manual. Dordrecht, The Netherlands: Kluwer Academic Publishers: 350–390.

Gill, H.K.; Capinera, J.L.; McSorley, R. 2008. Lesser cornstalk borer, *Elasmopalpus lignosellus* (Zeller) (Insecta: Lepidoptera: Pyralidae). Publication EENY155. Gainsville, FL: University of Florida, Institute of Food and Agricultural Sciences, Entomology and Nematology Department. 8 p.

Leuck, D.B. 1966. Biology of the lesser cornstalk borer in south Georgia. Journal of Economic Entomology. 59: 797–801.

Conifer and Hardwood Insects

47. Mole Crickets
Coleman Doggett

Hosts

Mole crickets are in the family Gryllotalpidae. The forelegs of these insects are modified for digging. Three introduced species—shortwinged mole cricket, *Scapteriscus abbreviatus*; southern mole cricket, *S. borellii*; and tawny mole cricket, *S. vicinus*—are very destructive to seedlings, while the several native U.S. species seldom cause damage. A wide variety of both hardwood and softwood species are killed by mole crickets, either by feeding or by the indirect effect of their tunneling. Seedlings attacked include the southern yellow pine, eastern redcedar, elm, maple, sweetgum, and yellow-poplar.

Distribution

The introduced species are currently found from Texas to New Jersey.

Damage

Mole crickets feed on seedling roots and lower stems. Belowground damage initially results in wilting and if damage is severe, eventual seedling death. The winding tunnels constructed by the insects cause topsoil disturbance, which may result in mortality to very small seedlings when rain splash covers them with soil. Seedlings may also be severed or girdled just above the groundline. When affected seedlings are examined, major and or minor roots may be missing due to insect feeding.

Diagnosis

Adult mole crickets are 2.6 to 5.2 cm (1.0 to 2.0 in) long with large shiny black eyes and range in color from yellowish brown to dark brown (fig 47.1). Their front legs are stout and shovel-shaped with claws for digging. They tunnel underground and spend most of their lives in their tunnels. The tunnels are around 3 mm in diameter and may be visible on the soil surface. Occasional small soil mounds on the ground surface above the tunnels indicate where the insects have surfaced.

Biology

Adult mole crickets lay 25 to 60 eggs in an underground chamber in April or May. One species, the shortwinged mole cricket, however, can produce eggs all year. The eggs hatch into nymphs in 3 to 4 weeks and the nymphs begin to build underground tunnels and feed. Nymphs are similar to adults except that their wings are not developed. They feed on plant roots and stems, earthworms, and other insects. There are one or two generations per year, with two generations occurring in the southern part of their range. After feeding for several months, the nymphs become adults, develop wings, fly to new locales, and construct tunnels. Adults are strong fliers and may fly up to 5 miles. Inside the tunnels, the males produce mating songs or rasps to attract mates. After mating, the female digs a nursery chamber and lays her eggs to begin a new generation. Mole crickets are nocturnal and are seldom seen in daylight.

Mole crickets have many natural enemies. Predators include birds, toads, shrews, moles, rats, skunks, raccoons, foxes, and armadillos.

Control

Biological

A parasitic wasp, *Larra bicolor*, was introduced from South America to control mole crickets. It is established in Florida and appears to be surviving and spreading. A parasitic fly, *Ormia deplete*, also from South America, has been introduced.

In addition, a nematode, *Steinemema scapterisci*, has been introduced and has been effective for control. It is commercially available and has proven to be persistent for up to 8 years.

Figure 47.1—*Mole cricket.* Photo by Richard Grantham, Oklahoma State University.

Conifer and Hardwood Insects

47. Mole Crickets

Chemical

A number of both general and restricted-use pesticides are currently registered for mole cricket control. These chemicals include both liquid and granular formulations, as well as baits. Chemical control is most effective when nymphs are small, late June and early July in the mid-South. Chemical control becomes less effective as the insects age.

Irrigation before insecticidal application may increase effectiveness by driving the insects closer to the surface. Some insecticides require post application irrigation to carry the insecticide to the mole crickets. These requirements are included on the product label.

Most available baits contain a grain-based attractant plus an insecticide. Since mole crickets are nocturnal, baits should be applied in the evening. The effectiveness of the baits that are currently available is decreased if rainfall or irrigation occurs before the insects feed on the bait.

Selected References

Anon. 1983. Insects of conifer roots: mole crickets. Bull. No. 196-A. Tallahassee, FL: Florida Division of Forestry. 1 p.

Arnold, D.; Rebeck, E.; Royer, T.; Mulder, P.; Kard, B. 2008. Major horticultural and household insects of Oklahoma. E-918. Oklahoma City: Oklahoma Cooperative Extension Service. 176 p.

Buss, E.A.; Capinera, J.L.; Leppla, N.C. 2006. Pest mole cricket management. Extension Publication ENY-324. Gainesville, FL: University of Florida, Institute of Food and Agriculture Sciences. 6 p.

USDA Forest Service. 1985. Insects of eastern forests. Misc. Pub. 1426. Washington, DC: USDA Forest Service. 608 p.

Conifer and Hardwood Insects

48. Plant Bugs
David B. South

Hosts

Plant bugs (*Lygus* spp.; *Taylorilygus apicalis*; family Miridae) feed on buds, flowers, and growing tips of plants. Seedlings of Douglas-fir, true fir, larch, pine, poplar, and spruce are commonly damaged by lygus bugs.

Distribution

Conifers are affected by plant bugs throughout the world. Nursery damage occurs in the Pacific Northwest (*L. hesperus, L. lineolaris,* and *L. elisus*), Canada, and North Central and Southern United States (*L. lineolaris*). *Taylorilygus apicalis* has a worldwide distribution and occurs on six continents.

Damage

Economic damage by plant bugs occurs mainly on 1-0 conifer seedlings, and feeding can severely affect the form of the seedling. When saliva is injected into the stem, it can cause needle twisting and forked seedlings (sometimes referred to as "multiple tops" or "bushy-tops"). If feeding occurs soon after germination, some seedlings will become stunted and these may end up as culls.

Diagnosis

Lygus and *Taylorilygus* are typically found feeding on weeds that produce white or yellow flowers (fig. 48.1). Monitoring the population can be done by close inspection of the flowers of daisy fleabane, cutleaf evening primrose, common groundsel, and many *Brassica* species. Use of traps (white or yellow sticky traps or clear plastic traps) also helps monitor plant bug population levels (fig. 48.2), but some managers choose to monitor weed levels because traps require time and money.

Adult *L. hesperus* (fig. 48.3) are yellowish-green to brown and the males are 5.3 to 6.5 mm long with a "V" marking on their back. Nymphs are 1 to 6 mm long, wingless, and appear similar to pale-green aphids. Adult *Taylorilygus* (fig. 48.4) are generally green. Most damage to 1-0 conifer seedlings occurs between May and September. Look for stem lesions and distorted needles. Feeding causes deformed or

Figure 48.1—*Example of a* Lygus lineolaris *on daisy fleabane.* Photo by Ronald Smith, Auburn University, at http://www.bugwood.org.

Figure 48.2—*Clear Plexiglas trap with a catch pan containing soapy water.* Photo by David B. South, Auburn University.

Conifer and Hardwood Insects

48. Plant Bugs

Figure 48.3—*Example of* Lygus hesperus. Photo by Whitney Cranshaw, Colorado State University, at http://www.bugwood.org.

Figure 48.4—*Example of* Taylorilygus apicalis. Photo copyright 2006 Bill Claff.

aborted buds and multiple tops (fig. 48.5). When present, some feeding also occurs in transplants and 2-0 seedlings. In 1- to 2-year-old hybrid poplars, feeding produces a split lesion in the middle to upper stem. Lesions result in gall formation and stems often break just above the wound.

Biology

Adults overwinter in plant debris along field edges and in transplant beds. In early spring, adults feed and lay eggs in stems of agricultural crops or herbaceous weeds. Within a few weeks, eggs hatch into flightless nymphs that, like adults, feed on plant juices. Three to four generations are completed per year in the Northern United States and five generations can occur in the South. Adults are active fliers and readily move from one crop to another. Irrigated and fertilized nursery crops apparently attract adults when nearby host plants mature, senesce, or are harvested.

Control

Netting is sometimes used in greenhouses, and some managers use netting to cover containers grown outside on benches. Netting has also reduced injury to 1-0 Fraser fir in seedbeds.

To avoid attracting plant bugs to the nursery, some managers keep weeds mowed to reduce the population of preferred weed species. Unfortunately, plant bugs can travel long distances and, therefore, additional control measures are needed. Some speculate that during dry periods, lygus bugs are attracted to succulent seedlings that have been irrigated and fertilized.

Injury to conifer seedlings from lygus bugs was rare in the decades when mineral spirits were applied for weed control. This situation might be due to the insecticidal properties of mineral spirits or because the volatile compounds produced an offensive odor. After managers ceased mineral spirit use, some began to notice an increase in the number of bushy-top seedlings. Because *Lygus* species have several generations per year, multiple insecticide applications may be necessary in outdoor nurseries (fig. 48.6). A few insecticides have proven effective in reducing the amount of damage to seedlings. In loblolly pine nurseries insecticide treatments commence as soon as *Lygus* adults are found feeding on weeds, which usually occurs in late April or early May.

Figure 48.5—*Multiple shoots resulting from feeding of plant bugs.* Photo by David B. South, Auburn University.

Conifer and Hardwood Insects

48. Plant Bugs

Figure 48.6—*The use of insecticides (right bed) can increase the crop value of sand pine seedlings.* Photo by Wayne N. Dixon, Florida Department of Agriculture and Consumer Services.

Selected References

Bryan, H. 1989. Control of the tarnished plant bug at the Carter's Nursery. Tree Planters' Notes. 40(4): 30–33.

Haseman, L. 1918. The tarnished plant bug and its injury to nursery stock (*Lygus pratensis* L.). Res. Bull. 29. Columbia, MD: University of Missouri, College of Agriculture, Agricultural Experiment Station.

Holopainen, J.K.; Rikala, R.; Kainulainen, P.; Oksanen, J. 1995. Resource partitioning to growth, storage and defense in nitrogen fertilized Scots pine and susceptibility of the seedlings to the tarnished plant bug *Lygus rugulipennis*. New Phytologist. 131: 521–532.

Holopainen, J.K.; Varis, A.L. 2009. Host plants of the European tarnished plant bug *Lygus rugulipennis* Poppius (Het., Miridae). Journal of Applied Entomology. 111(1–5): 484–498.

Kohmann, K. 2008. Multiple leaders caused by the tarnished plant bug (*Lygus rugulipennis*) in *Picea abies* seedlings. Scandinavian Journal of Forest Research. 21(3): 196–200.

Overhulser, D.L.; Kanaskie, A. 1989. Lygus bugs. In: Cordell, C.E.; Anderson, R.L.; Hoffard, W.H.; Landis, T.D.; Smith, Jr., R.S.; Toko, H.V., tech. coords. Forest nursery pests. Agriculture Handbook 680. Washington, DC: USDA Forest Service: 146–147.

Sapio, F.J.; Wilson, L.F.; Ostry, M.F. 1982. A split-stem lesion on young hybrid Populus trees caused by the tarnished plant bug, *Lygus lineolaris* (Hemiptera (Heteroptera: Miridae)). The Great Lakes Entomologist. 15(4): 237–246.

Schowalter, T.D.; Overhulser, D.L.; Kanaskie, A. [and others]. 1986. *Lygus hesperus* as an agent of apical bud abortion in Douglas-fir nurseries in western Oregon. New Forests. 1: 5–15.

Schowalter, T.D.; Stein, J.D. 1987. Influence of Douglas-fir seedling provenance and proximity to insect population sources on susceptibility to *Lygus hesperus* (Heteroptera: Miridae) in a forest nursery in western Oregon. Environmental Entomology. 16(4): 984–986.

South, D.B. 1991. Lygus bugs: a worldwide problem in conifer nurseries. In: Sutherland, J.R.; Glover, S.G., eds. Proceedings, 1st IUFRO Workshop on Diseases and Insects in Forest Nurseries. Info. Rep. BC-X-331. Victoria, British Columbia, Canada: Forestry Canada: 215–222.

South, D.B.; Zwolinski, J.B.; Bryan, H.W. 1993. *Taylorilygus pallidulus* (Blanchard): a potential pest of pine seedlings. Tree Planters' Notes. 44(2): 63–67.

Wheeler, A.G. 2001. Biology of the plant bugs (Hemiptera: Miridae): pests, predators, opportunists. Ithaca, NY: Comstock Publishing Associates. 528 p.

Wheeler, Jr., A.G.; Henry, T.J. 1992. A synthesis of the Holarctic Miridae (Heteroptera): distribution, biology, and origin, with emphasis on North America. Lanham, MD: The Thomas Say Foundation, Entomological Society of America. 282 p.

Wilson, L.F.; Moore, L.M. 1985. Vulnerability of hybrid Populus nursery stock to injury by the tarnished plant bug. The Great Lakes Entomologist. 18: 19–23.

Conifer and Hardwood Insects

49. Seed and Cone Insects
Alex C. Mangini

Hosts

Seed and cone insects are not nursery pests in the strict sense; they are seldom found on nursery trees. They are important because they destroy seed available for planting in nurseries. These insects are mostly conifer pests, but some affect hardwoods. Most conifer nurseries rely on seed grown in managed seed orchards that are located in the South and Pacific Northwest; some are in the upper Midwest. Insects damage acorn crops in Southern oak orchards.

The major conifer seed and cone insects are the coneworms, *Dioryctria* species (fig.49.1); cone beetles, *Conophthorus* species (fig. 49.2); seed bugs, *Leptoglossus* species (fig. 49.3); seed worms, *Cydia* species; gall midges, *Contarinia* species; cone borers, *Eucosma* species; and cone moths, *Barbara* species. Weevils, *Curculio* species (fig. 49.4), and the filbertworm, *Cydia latiferranea*, attack oak acorns and other hardwood seeds.

In the South, loblolly, longleaf, slash, and shortleaf pines are hosts. The primary pests are the leaffooted pine seed bug, *L. corculus,* and five coneworm species. Additional pests include the shieldbacked pine seed bug, *Tetyra bipunctata*, and several seed worm species. Slash pine is also attacked by the slash pine flower thrips, *Gnophothrips fuscus*. A serious eastern white pine pest is *Conophthorus coniperda*, the white pine cone beetle.

In the Western States, Douglas-fir, western hemlock, western redcedar, and western larch are hosts, as are ponderosa, lodgepole, sugar, western white, and other pines. The western conifer seed bug, *L. occidentalis*, is a Douglas-fir, pine, and grand fir pest. Several coneworm species attack Douglas-firs, true firs, and pines. The Douglas-fir cone gall midge, *Contarinia oregonensis*, is a serious pest in Douglas-fir seed orchards, along with the Douglas-fir seed chalcid, *Megastigmus spermotrophus*.

Figure 49.2—*A female cone beetle,* Conophthorus *species, excavating a gallery along the axis of a developing pine cone. She will deposit eggs along the gallery.* Photo by Steven Katovich, USDA Forest Service, at http://www.bugwood.org.

Figure 49.3—*An adult female leaffooted pine seed bug,* Leptoglossus corculus, *on pine needles.* Photo by R. Scott Cameron, Advanced Forest Protection, Inc., at http://www.bugwood.org.

Figure 49.1—*Larva and adult of the southern pine coneworm,* Dioryctria amatella, *on an infested loblolly pine cone. Larval feeding has partially destroyed the cone.* Photo by Larry R. Barber, USDA Forest Service, at http://www.bugwood.org.

Conifer and Hardwood Insects

49. Seed and Cone Insects

Figure 49.4—*An adult female curculio weevil,* Curculio *species, on a young acorn. She will use her elongated mouthparts to excavate a hole in the acorn and deposit eggs.* Photo by Larry R. Barber, USDA Forest Service, at http://www.bugwood.org.

Distribution

Seed and cone insects are found wherever their host tree species occur. Some species are host-specific such as the Douglas-fir cone gall midge and Douglas-fir seed chalcid. Other species have a wide geographical distribution and feed on multiple hosts; for example, the western conifer seed bug.

Seed and cone insects can be divided into two geographic groups, those present in the Western and Northern States on pine, spruce, fir, and Douglas-fir and those found in the Southern and Atlantic-Coastal States, primarily on pines. For example, *L. occidentalis* is found in the West and *L. corculus* (fig. 49.3) inhabits the Southern States. Both species are similar in appearance and behavior but have disjunct distributions. Similarly, distinct coneworm species groups occur in the West and South.

The filbertworm is found throughout the country on various oak and hickory species. Several curculio weevil species are present across the country.

Damage

The damage is a lost or reduced seed crop. Damage occurs on either the cones, the seeds, or both. Most damage is done by the larval stages, which is particularly true for cone beetles, coneworms, seed worms, and cone borers. Some species feed on the developing cone and incidentally destroy seeds. Some species feed on cone and seed tissues as they move from seed to seed. Other pests develop entirely within a seed. Damage is either external or internal.

External damage includes dead or dying cones (fig. 49.5), dead or damaged flowers and conelets, galls, abnormal development, and feeding damage. Frass and webbing may be present. Male flowers can be damaged or deformed. Internal damage can be seen when the cone or conelet is cut or broken apart. Internal damage includes dead tissue, frass, or galls. Frass (solid larval excrement) and webbing may be present; the larva(e) may often be seen. Damaged seeds may be discolored, shrunken, or stuck to galls. Internally, seeds may be empty, or display shrunken or dead contents.

Diagnosis

Externally, look for dead and dying cones, discolored cones, dead flowers, and conelets and distorted or abnormal cones. Keep in mind the host species biology; insect life cycles are tied to host phenology, and damage can occur at specific cone development stages. Damage may occur in more than one host stage; for example, coneworms can damage both first-year and second-year cones in pines.

Figure 49.5—*Damage caused by* Conophthorus ponderosae, *a cone beetle. The white pitch tube at the bases of the cones indicates where the attacking female entered.* Photo from USDA Forest Service, Ogden Archive, at http://www.bugwood.org.

Conifer and Hardwood Insects

49. Seed and Cone Insects

Look for specific indicators. Coneworms are indicated by frass and webbing presence on damaged cones or conelets (fig. 49.6). In southern pines, coneworm-killed conelets break apart easily to reveal frass and webbing. When cones occur in clusters, the coneworms will progressively kill each cone in the cluster. Cone borer damage is similar to coneworm damage but frass is more tightly packed. Cone moth damage is also similar to coneworm damage but pitch is more abundant in the frass. Cone beetles usually enter at the cone base; a pitch tube is often present. For the Douglas-fir cone gall midge, cone scales die and turn red in July or August.

To diagnose internal damage, one must cut open the cone, seed, or acorn. In some instances, seed can be radiographed to assess damage. Again, look for specific indicators. Seed bug damage results in empty or partially filled seeds. Empty seeds show up clearly in radiographs (fig. 49.7). Similarly, Douglas-fir seed chalcid larvae can be seen within seeds on radiographs. Also, a distinct exit hole is made by the emerging chalcid wasp. Cone beetles can be found in cones and appear as white, legless, C-shaped larvae with brown heads or as distinctive pupae. The filbertworm appears within acorns as a reddish larva surrounded by coarse frass and webbing. Curculio weevils appear as legless, C-shaped larvae with densely packed frass.

Figure 49.6—*A larva of the webbing cone worm,* Dioryctria disclusa, *that has been feeding on a pine conelet. Note the frass and webbing at the base of the conelet.* Photo by Larry R. Barber, USDA Forest Service, at http://www.bugwood.org.

Biology

Seed and cone insect biology is closely tied to the reproductive biology of the host tree. In general, larvae feed (and cause damage); adults disperse; the stage that overwinters is variable.

Coneworm adults are small moths with distinctive crossbands on wings. Females attract males with species-specific pheromones. Coneworm eggs are laid on branches. Several species overwinter as early instar larvae that feed externally before entering flowers, shoots, and cones. Larvae are elongate and yellow to purple in color with dark head capsules. As the larvae feed, they kill the cone and its seeds. Pupation is in the killed cones. Some species have only one generation per season (univoltine) and some have multiple generations (multivoltine).

Figure 49.7—*A radiograph of longleaf pine seeds. Healthy seeds are white with visible embryo. Seeds damaged by seed bugs have shriveled contents; seeds damaged by seed worm have gray contents and broken seed coats.* Photo by Alex C. Mangini, USDA Forest Service.

Conifer and Hardwood Insects

49. Seed and Cone Insects

Cone beetles are dark brown to black, shiny, stout beetles that are related to bark beetles. The female attacks the cone at the base and makes a tunnel along the cone axis. She deposits eggs along the gallery. Eggs are shiny and ovoid. The larvae are white, with brown heads, and feed on the developing cone, destroying the seeds. After pupation, new adults either overwinter in cones or enter pine shoots until spring. There is one generation per year.

Seed bug adults are elongate, 18 to 19 mm long, reddish brown with a distinctive zigzag pattern on the forewing (fig. 49.3). Seed bugs do not have larvae or pupae. Their immatures, called nymphs, are similar to adults but are smaller and without fully developed wings (fig. 49.8). The eggs are barrel-shaped and laid end-to-end on the needles. The western conifer seed bug and the leaffooted pine seed bug have similar life cycles, except that the former has only one generation per year while the latter may have several. Adults overwinter and become active in spring, feeding initially on male flowers. Eggs are laid in spring. Five nymphal instars exist. The second-instar leaffooted pine seed bug nymphs feed on conelet ovules causing conelet abortion (fig. 49.9). Later instars and adults feed on seeds in the developing cones leaving the seeds empty.

Douglas-fir cone gall midge adults are small, only 3 to 4 mm long. They emerge in early spring. The females oviposit near young ovules when Douglas-fir flowers are open for pollination. The eggs hatch in May and June. The tiny orange larvae tunnel into scale tissue near ovules and develop through three instars; during this development a gall forms around the larva. In the fall, the larvae drop to the ground and form an overwintering cocoon—often in old male flowers. Seeds may become fused to the midge galls; consequently the seed are not shed from the mature cone (fig. 49.10).

Figure 49.8—*A second-instar nymph of the leaffooted pine seed bug,* Leptoglossus corculus, *feeding on a pine conelet. Conelets often abort after the nymphs feed on them.* Photo by Tim Tigner, Virginia Department of Forestry, at http://www.bugwood.org.

Figure 49.9—*Ovules dissected from ponderosa pine conelet. The top ovule has been fed on by the western conifer seed bug and appears brown and shrunken. The bottom ovule is healthy.* Photo by Alex C. Mangini, USDA Forest Service.

Control

Biological control of seed and cone insects has not been fully explored. Cultural control consists of orchard sanitation. For example, white pine cone beetle populations can be reduced by running a controlled burn through the orchard in late fall or early spring. The beetles overwinter within the killed cones that drop to the ground. Burning these cones kills the overwintering beetles. Similarly, the Douglas-fir cone gall midge overwinters on the ground. It has been demonstrated that removing the duff from the orchard floor reduces the impact of this insect.

Seed orchard managers use integrated pest management (IPM) to optimize pest control efforts; all available tools are used—including cultural and chemical control. Chemical control of seed and

Conifer and Hardwood Insects

49. Seed and Cone Insects

Figure 49.10—*Douglas-fir cones infested by larvae of the Douglas-fir cone gall midge,* Contarinia oregonensis. *The larvae cause galls to form on cone scales. Heavy infestations kill the cones.* Photo by Ward B. Strong, British Columbia Ministry of Forests, British Columbia, Canada.

cone insects is difficult because these species spend most of their life cycle within the conelet or cone sheltered from pesticide sprays. Managers must time applications appropriately and ensure good coverage. Research has led to degree-day models for the southern pine coneworm; sprays can be timed to coincide with early instar larvae presence before they enter cones. Using different pesticide chemistries during the season, avoids resistance and secondary pests. New pesticides, such as growth regulators, are safer, more specific, and have less nontarget impact.

Selected References

Barber, L.R. 1989. Seed and cone insects. In: Cordell, C.E.; Anderson, R.L.; Hoffard, W.H.; Landis, T.D.; Smith, Jr., R.S.; Toko, H.V., tech. coords. Forest nursery pests. Agriculture Handbook 680. Washington, DC: USDA Forest Service: 84–85.

Ebel, B.H.; Flavell, T.H.; Drake, L.E.; Yates III, H.O.; DeBarr, G.L. 1980. Seed and cone insects of southern pines. GTR-SE-8, rev. Asheville, NC: USDA Forest Service, Southeast Area, State and Private Forest; Atlanta, GA: USDA Forest Service, Southeast Forest Experimental Station. 43 p.

Furniss, R.L.; Carolin, V.M. 1977. Western forest insects. Misc. Publ. 1339. Washington, DC: USDA Forest Service. 654 p.

Gibson, L.P. 1982. Insects that damage northern red oak acorns. Res. Pap. NE-492. Broomall, PA: USDA Forest Service, Northeast Forest Experimental Station. 6 p.

Hagle, S.K.; Gibson, K.E.; Tunnock, S. 2003. A field guide to the diseases and insect pests of northern and central Rocky Mountain conifers. Rep. R1-03-08. Ogden, UT: USDA Forest Service, Intermountain Region. 197 p.

Hedlin, A.F.; Yates III, H.O.; Cibrian Tovar, D.; Ebel, B.H.; Koerber, T.W.; Merkel, E.P. 1981. Cone and seed insects of North American conifers. Ottawa, Canada: Canadian Forestry Service: 122 p.

Conifer and Hardwood Insects

50. Mites
Alex C. Mangini

Hosts

The spider mites are the primary mite pests in forest nurseries. A second group, the rust and gall mites, can cause significant damage.

Spider mites (family Tetranychidae) in the genus *Oligonychus* are often pests on conifers. The most important species in this group is the spruce spider mite, *Oligonychus ununguis*, which attacks spruce, hemlock, fir, juniper, larch, redwood, incense cedar, and pine species. *O. milleri* can cause damage to nursery seedlings. *O. subnudus* and *O. coniferarum* infest conifers and are common in the Western States. *O. ilicis* is found on conifers in the Southern States. Some *Oligonychus* attack deciduous trees. *O. bicolor* is found on oaks and *O. platani* on sycamores and related species. Spider mite species in the genus *Eotetranychus* attack a variety of hardwoods.

The rust and gall mites (superfamily Eriophyoidea) cause two kinds of damage. Rust mites feed on the leaf surface and cause discoloration; gall mites cause the host plant to make distinctive growths on leaves and buds. The mites live in these galls. Rust mites in the genus *Trisetacus* are the eriophyoids most likely to be encountered in nurseries. Numerous species infest pine, spruce, cedar, Douglas-fir, cypress, and juniper.

Distribution

Spider mites and rust and gall mites are widespread. They can be found across the range of their hosts. Some mite species are host-specific.

Damage

Reduced tree growth and vigor are the major effects of severe mite infestations. Seedlings and small trees may be weakened and made susceptible to other problems. Trees are seldom killed by mites; however, seedling mortality may occur. Feeding damage makes trees unsightly.

Spider mites use their needle-like mouthparts (fig. 50.1) to pierce plant cells and suck out the cell contents, resulting in yellowing or browning of needles and leaves (fig. 50.2). Feeding by the spruce spider mite results in a mottled needle appearance; yellow spots appear on needles as the mites feed. The needles eventually turn yellow or brown. Associated with the discolored needles is a dense webbing made by the mites (fig. 50.3).

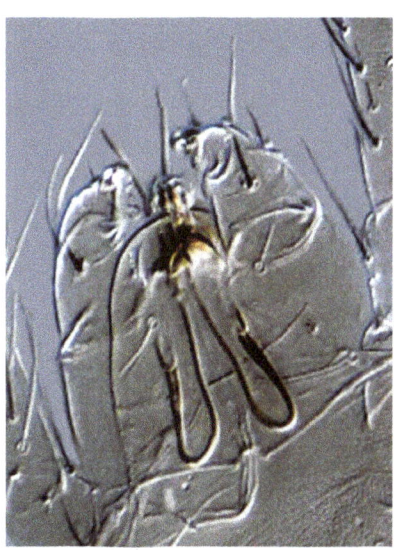

Figure 50.1—*Magnified (200x) image of a spider mite,* Tetranychus platani, *showing the mouthparts. The mite uses the long, paired, recurved stylets to pierce the plant cell and extract the cell contents.* Photo by Alex C. Mangini, USDA Forest Service.

Figure 50.2—*Browned needles resulting from an infestation of the spruce spider mite,* Oligonychus ununguis, *a serious pest in conifer nurseries and plantations.* Photo from USDA Forest Service—Northeastern Area Archives, at http://www.bugwood.org.

Conifer and Hardwood Insects

50. Mites

Figure 50.3—*Spruce spider mites and the webbing they produce are visible in this branch tip from an infested conifer.* Photo from USDA Forest Service—Region 4 Archive, at http://www.bugwood.org.

Figure 50.4—*Infestation of a white pine needle sheath by rust mites* (Trisetacus). *The top needle shows feeding damage. Mites are visible in the center right.* Photo by Alex C. Mangini, USDA Forest Service.

Rust mites cause similar damage to that caused by the spruce spider mite. *Trisetacus* species on conifers discolor needles (fig. 50.4). When infestations are severe, the needles yellow and eventually turn a reddish-brown or rusty color (hence the name rust mites). Feeding by gall-making eriophyoid mites causes the host to form distinctive, sometimes bizarre shaped galls (fig. 50.5). These gall-formers are common on deciduous hardwood hosts. As with the spider mites, the eriophyoids have piercing mouthparts that enable them to penetrate the plant cells.

Figure 50.5—*Spindle-shaped leaf galls caused by gall mites on a* Prunus *leaf. The mites live and feed within the galls.* Photo by Steven Katovich, USDA Forest Service, at http://www.bugwood.org.

Conifer and Hardwood Insects

50. Mites

Diagnosis

Spider mite presence is indicated by leaves or needles that are pale, washed-out, and discolored—yellow or brown needles are common (fig. 50.6). Heavy infestations during hot, dry summers can cause defoliation. Most, but not all, spider mite species form a fine webbing on needles and leaves. For example, spruce spider mites produce webbing that can form a dense covering over needles. Mites can be seen by shaking an infested branch over a white paper sheet. The mites fall onto the paper and appear as tiny red, yellow, or green specks on the paper. A careful look using a hand lens will show the mites walking on the paper. The mites have globular bodies (fig. 50.7) with four pairs of legs (three pairs in the larval stage). The eggs may be spherical or spherical with a stipe—a thin extension at the top. For example, spruce spider mite eggs are green when laid and turn reddish brown (fig. 50.7).

Eriophyoid rust mite infestations are indicated by leaves that yellow and eventually turn brown or rusty. Rust mites are extremely small—0.1 to 0.5 mm long—and must be seen using a hand lens or microscope. They are elongated and worm-like with four legs at the head end of the body. They are yellowish-white in color and are slow moving (fig. 50.8).

Rust mite damage on conifers is similar to that caused by spider mites. To distinguish the damage, shake an infested limb over a white paper sheet. Spider mites will be visible on the paper as dark specks. Rust mites, however, are far too small to be seen on the paper without a hand lens. The presence of webbing will also distinguish the spruce spider mite from rust mites. Trees with bronzing should be examined by removing the needle sheath and using a hand lens to examine the exposed needle bases.

Biology

Spider mites have five life stages—egg, larva, protonymph, deutonymph, and adult. The larva has three pairs of legs; the nymph and adult stages have four. Spider mites have very short life cycles and populations can grow rapidly. The spruce spider mite overwinters as eggs at the needle base. The eggs hatch in spring and reach the adult stage in about 15 days. Several generations occur in a season. One female spider mite can lay up to 50 eggs. Most eggs develop into females. As a result, populations expand rapidly when conditions are good. The mites and their eggs are protected by the dense webbing they produce. The mites disperse by wind; because of their small size, they are easily lifted into the air on wind currents.

Rust mites in the genus *Trisetacus* overwinter as adults and eggs in the needle sheaths of their conifer hosts. In the spring, they move to new growth. Several

Figure 50.6—*Ponderosa pine seedlings damaged by* Oligonychus subnudus *show the characteristic pale, yellow appearance resulting from extensive spider mite feeding.* Photo by Whitney Cranshaw, Colorado State University, at http://www.bugwood.org.

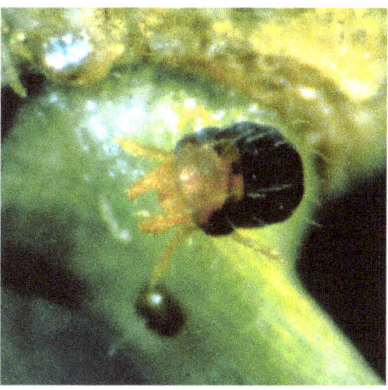

Figure 50.7—*An adult female of the spruce spider mite,* Oligonychus ununguis. *The body is dark while the legs and mouthparts are lighter in color. Below the female is an egg with the characteristic thin stipe at the top of the egg.* Photo from USDA Forest Service—Northeastern Area Archives, at http://www.bugwood.org.

Conifer and Hardwood Insects

50. Mites

overlapping generations occur per year. When populations become very large, the mites will leave the needle sheaths and feed on the exposed needle surface. These mites are usually present on their conifer hosts at all times. Population outbreaks occur when weather conditions are favorable. Also, repeated insecticide application promotes outbreaks by eliminating natural enemies.

Control

Biological

Both spider mites and rust mites have natural enemies that keep their populations in check. Several arthropod natural enemies feed on spider mites. These enemies include beetles in the family Coccinellidae, anthocorid bugs, predaceous thrips, and predaceous mites in the family Phytoseiidae. By far the most important biological control agents are the phytoseiid mites.

Phytoseiid mites are common associates of spider mites and have been much studied. About 2,000 species exist worldwide. Most species are predatory but also feed on pollen, honeydew, and rust mites. Many are specialist spider mite predators (fig. 50.9); these include species in the genera *Phytoseiulus*, *Galendromus,* and *Neoseiulus*. Phytoseiids have a short life cycle, produce many eggs per female, and have a preponderance of females in the population. Under ideal conditions, a 5-day generation time is common; one female can lay up to five eggs per day for several weeks. This rapid population growth makes them useful control agents against spider mites. Phytoseiid mites are available commercially for use in greenhouses and nurseries.

Cultural

Mite infestations can build up quickly. During the growing season, trees should be monitored for mite presence. If found, mites and eggs can be washed off with a strong water spray. Proper irrigation to maintain growth will reduce the impact of mites. Avoid planting host species such as pine, cedar, hemlock, fir, juniper, or spruce adjacent to windbreaks of the same species.

Chemical

Spider mites and rust mites may occur in nurseries after insecticide application that controls other pests. The natural mite enemies are killed and conventional insecticides may not affect mites. It may be necessary to apply miticides in severe infestations. Modern miticides are the avermectins, organotins, and the benzoylureas. The latter interrupt growth by inhibiting transition from one stage to the next. Products should be alternated to avoid pesticide resistance in mites. When applying miticides, it is important to ensure thorough coverage of the plants.

Selected References

Furniss, R.L.; Carolin, V.M. 1977. Western forest insects. Misc. Publ. 1339. Washington, DC: USDA Forest Service. 654 p.

Goheen, E.M.; Willhite, E.A. 2006. Field guide to common diseases and insect pests of Oregon and Washington conifers. R6-NR-FID-PR-01-06. Portland, OR: USDA Forest Service, Pacific Northwest Region. 327 p.

Jeppson, L.R.; Keifer, H.H.; Baker, E.W. 1975. Mites injurious to economic plants. Berkeley, CA: University of California Press. 614 p.

Johnson, W.T.; Lyon, H.H. 1991. Insects that feed on trees and shrubs, 2nd ed. Ithaca, NY: Cornell University Press. 560 p.

Walter, D.E.; Proctor, H.E. 1999. Mites: ecology evolution and behaviour. Wallingford, Oxon, United Kingdom: CAB International. 322 p.

Figure 50.8—*An adult* Trisetacus *rust mite feeding on a white pine needle. Note the elongated body and the two pairs of legs. The mouthparts are between the first pair of legs.* Photo by Alex C. Mangini, USDA Forest Service.

Figure 50.9—*A pale yellow phytoseiid mite,* Galendromus occidentalis, *feeding on a European red mite,* Panonychus ulmi. *Phytoseiids are important biological control agents against spider mites.* Photo by Elizabeth H. Beers, Washington State University.

Conifer and Hardwood Insects

51. Weevils
Art Antonelli

Revised from chapter by William M. Hoffard, 1989.

Hosts

Weevils (family Curculionidae) feed on a wide variety of plants. In forest nurseries, seedlings of many species, especially hemlock, spruce, and Douglas-fir, may be severely attacked. The most destructive weevils in nurseries are members of the genus *Otiorhynchus* (syn. *Brachyrhinus*). These weevils, collectively referred to as root weevils, include the strawberry root weevil, *O. ovatus*; the black vine weevil, *O. sulcatus*; the rough strawberry weevil, *O. rugusostriatus*; and the woods weevil, *Nemocastes incomptus*. Other weevil species that can cause problems in nurseries are the pales weevil, *Hylobius pales*; the pitch-eating weevil, *Pachylobius picivorus*; the Japanese weevil, *Pseudocneorrhinus bifasciatus*; and the yellow-poplar weevil, *Odontopus calceatus*.

Distribution

The most severe losses due to weevils have been observed in the West, but damage can occur almost anywhere.

Damage

Weevils sometimes cause serious damage to various species of conifer seedlings, especially in the West where they are among the most serious insect nursery pests. Outright mortality may be widespread. Heavy culling is sometimes necessary due to severe damage.

Diagnosis

On aboveground plant parts, look for needles with notches or holes (fig. 51.1) or stems with small sections of bark removed. Infested roots or crowns may be entirely stripped of their bark for several centimeters (fig. 51.2) or debarked on only one side. Root-damaged trees may show symptoms reminiscent of damage due to root pathogens. Adult weevils have well-developed snouts with clubbed, elbowed antennae (figs. 51.3 and 51.4). Adults can be found in the soil litter layer during the day and on foliage at night. (Some species estivate in the soil during hot summer months and become active in the fall.) Larvae are legless and C-shaped, with shiny smooth heads, and pale cylindrical bodies (fig. 51.5).

Figure 51.1—*Notches on needles made by feeding of adult weevils.* Photo by Chal Landgren, Oregon State University Extension Service.

Biology

The life histories of weevils vary according to species and geographic influences. A generalized life history of the root weevils mentioned previously will serve as an example. Weevils overwinter as larvae or sometimes as pupae and adults at soil

Figure 51.2—*Bark stripped from the lower stem and roots of a fir seedling by feeding of weevil larvae.* Photo by Thomas D. Landis, USDA Forest Service.

Conifer and Hardwood Insects

51. Weevils

Figure 51.3—*Adult black vine weevil.* Photo by Ken Gray. Image courtesy of Oregon State University.

Figure 51.4—*Adult obscure root weevil.* Photo by Ken Gray. Image courtesy of Oregon State University.

depths of 15 to 20 cm (6 to 8 in). They emerge as adults from early May to late July. These adults are present throughout the remainder of the growing season. Sometimes two weevil generations are born per year, but one is more common. The weevils are night feeders, preferring to hide beneath leaves, stubs, stones, and other debris during the day. At night, they become active, climbing seedling stems and eating notches in needles and stems. Egg-laying begins about 4 to 6 weeks after adults appear and can continue until September, depending on species. In about 20 days, the eggs hatch, and the tiny larvae move deeper into the soil, where they feed first on fine roots and later on larger roots. Pupation occurs in cells just beneath the soil surface.

Monitoring

Determining the onset of adult weevil emergence and feeding is important in maximizing egg-laying suppression since newly emerged females do not begin to lay eggs for 30 days or more. If effective control methods are implemented within 1 month after emergence, it is possible to reduce the local weevil population by nearly 100 percent. Several monitoring techniques can be employed to determine the onset of weevil emergence. New leaves can be observed in early spring for notching. In addition, the presence of adults can be confirmed at night in early spring with a flashlight by inspecting foliage of strategically placed bait or trap crops, such as susceptible, nonresistant rhododendrons on warm, still evenings. Trapping is still another technique worthy of discussion. Laying a small piece of cardboard at the base of affected plants provides a hiding place that can be checked the next morning. Pitfall traps buried near the plants will achieve the same end.

Control

Cultural

Practicing clean cultivation, rotating transplant beds, and allowing infested areas to lie fallow and be thoroughly cultivated in alternate years are effective methods of reducing weevil populations.

Chemical

Fumigate seedbeds with a registered product before seeds are sown. Fumigation is very effective in controlling soil-inhabiting insects like weevils. Application of a registered adulticide may be advisable. Make sure application occurs before egg-laying begins.

Figure 51.5—*Root weevil larvae.* Photo by Art Antonelli, Washington State University.

Conifer and Hardwood Insects

51. Weevils

Biological

Entomophagous nematodes can control weevils very effectively in container-grown crops. These creatures contain lethal bacteria that are released into the body cavities of larvae and pupae. There are several caveats that need to be adhered to if this technique is to be effective. First, inspect the product for quality and morbidity to be certain the nematodes are mostly alive at time of purchase. Also, thoroughly wet the soil prior to application so that the nematodes can move through the soil particles easily in search of larvae and to maintain their own body moisture. Never apply the nematodes in direct sunlight since ultraviolet light will kill them. The best time for application is probably in the morning before the sun comes up or in the evening as the sun is setting. The soil temperature must be 11 °C (52 °F) or warmer at the time of application to activate the nematodes. Finally, if you apply nematodes with a sprayer that has a filtering mesh screen in the nozzle be sure to remove it because it will stop movement of the nematodes through the nozzle and onto the soil.

Selected References

Chastagner, G.A.; Byther, R.S.; Antonelli, A.; DeAngelis, J.; Landgren, C., eds. 1997. Christmas tree diseases, insects, and disorders in the Pacific Northwest: identification and management. Washington State University Extension Miscellaneous 0186. Pullman, WA: Washington State University Extension. 154 p.

Furniss, R.L.; Carolin, V.M. 1977. Western forest insects. Misc. Pub. 1339. Washington, DC: USDA Forest Service. 654 p.

Hoffard, W.H. 1989. Weevils. In: Cordell, C.E.; Anderson, R.L.; Hoffard, W.H.; Landis, T.D.; Smith, Jr., R.S.; Toko, H.V., tech. coords. Forest nursery pests. Agriculture Handbook 680. Washington, DC: USDA Forest Service: 142–143.

Hollingsworth, C.S.; Antonelli, A.; Hirnyck, R., eds. 2010. Pacific Northwest insect management handbook. Corvallis, OR: Oregon State University Press. 698 p.

Smith, F.F. 1932. Biology and control of the black vine weevil. USDA Tech. Bull. 325. Washington, DC: USDA Forest Service. 45 p.

USDA Forest Service. 1985. Insects of eastern forests. Misc. Publ. No. 1426. Washington, DC: USDA Forest Service. 608 p.

Conifer and Hardwood Insects

52. White Grubs
Albert E. Mayfield, III

Hosts

White grubs are soil-dwelling larvae of insects commonly known as "May beetles" or "June beetles" in the family Scarabaeidae. These grubs feed on herbaceous plant roots and other soil organic matter, but will also feed on the roots of woody plants, including all types of coniferous and hardwood seedlings in nursery settings. Numerous genera known to cause damage in forest nurseries include the common *Phyllophaga* (with more than 100 different species), *Polyphylla*, *Diplotaxis*, *Dichelonyx*, *Serica*, *Cotalpa*, *Anomala*, and others.

Distribution

Phyllophaga species and other white grubs are widely distributed, and can be found throughout much of North America, although the geographic range of an individual species may be more restricted.

Damage

Depending on the severity and extent of root injury, damage by white grubs kills seedlings or reduces their growth and vigor. Substantial losses may occur when more than one grub per 0.1 m² (1 ft²) of soil surface exists. When white grub problems occur, they may be more severe on light (i.e., sandy) soils and in newly turned seedbeds. White grub populations can become abundant beneath dense sod, weeds, or agricultural and cover crops in fallow years, thus nursery bed establishment on or near these sites may increase the risk of damage.

Diagnosis

White grub damage is typically noticed from June through early fall, when formerly healthy seedlings become discolored, wilt, and die. Aboveground symptoms may resemble drought injury. Heavily damaged seedlings can be pulled gently from the soil due to extensive root loss (fig. 52.1) Belowground, the taproot or lateral roots may be chewed off, girdled, gouged or debarked (fig. 52.2).

White grub larvae are fairly large and distinct, and may be readily detected in freshly prepared or turned soil. Larval size varies with age and species, but mature larvae are typically 20 to 45 mm (0.8 to 1.8 in) long, C-shaped, creamy-white, with a brown head, prominent mouthparts, and three pairs of well-developed legs (figs. 52.3 and 52.4). The abdomen is usually slightly enlarged and translucent, allowing internal contents to be seen through the skin.

Figure 52.1—*Nursery-grown pine seedlings, showing root consumption by white grubs.* Photo from Florida Department of Agriculture and Consumer Services, Division of Forestry.

Conifer and Hardwood Insects

52. White Grubs

Figure 52.2—*Extensive white grub damage to seedling on left, including severed taproot, missing lateral roots, and debarked surfaces.* Photo by James D. Solomon, USDA Forest Service, at http://www.bugwood.org.

Adults in the genus *Phyllophaga* are robust, oval-shaped beetles with prominent legs, commonly 12 to 25 mm long, varying in color from yellow to reddish-brown to black (fig. 52.5). Other May and June beetle species may be shiny and brightly colored. Adults are nocturnal and not necessarily evident at the site where larvae have caused seedling damage.

Biology

The life cycle of *Phyllophaga* species may require 1 to 4 years to complete, depending on species and geographic location. Species with 2- and 3-year life cycles are common, and due to multiple species and overlapping broods, all life stages may occur during any given year at a particular location.

Adults emerge from the ground in the evenings (typically in May and June) and move to the foliage of nearby trees or other vegetation to mate and feed. At dawn, mated females return the soil and lay eggs at depths of 3 to 20 cm (1.2 to 8.0 in) beneath the surface. Larvae hatch within 2 to 3 weeks and begin to feed, first on soil organic matter and eventually on nearby seedling roots. In autumn, larvae migrate downward to depths of up to 1.5 m (5 ft), depending on temperature, frost levels, and soil characteristics, and remain inactive until spring, when they return to near the soil surface to feed again on roots. Larvae may repeat these patterns of root-feeding in the warm season and downward migration to overwinter for 1 or more years. When larvae are fully mature they pupate in earthen cells and later emerge as adults.

Conifer and Hardwood Insects

52. White Grubs

Figure 52.3—*White grub larvae.* Photo from Clemson University—USDA Cooperative Extension Slide Series, at http://www.bugwood.org.

or infested planting ground, particularly when grubs are nearest the surface (late spring through early fall), may help destroy or reduce grub populations.

Chemical

Seedbed fumigation can eliminate white grubs in the upper soil horizons, but overwintering larvae may reside or migrate to depths below the effective fumigation zone. Granular and liquid formulations of insecticides may also be used against white grubs. Irrigating the soil before and after insecticide application may help to bring grubs nearer to the surface and move insecticide into the soil, respectively.

Control

Cultural

Identify potential problem areas by scouting for white grubs in the soil at the start of and throughout seedbed preparation. Repeated disking of new, fallow,

Figure 52.4—*Three different species of white grub larvae, illustrating size variation (left to right:* Popillia japonica, Amphimallon majalis, *and* Phyllophaga *species).* Photo by David Cappaert, Michigan State University, at http://www.bugwood.org.

Figure 52.5—*Adult June beetle* (Phyllophaga *species*). Photo by Steven Katovich, USDA Forest Service, at http://www.bugwood.org.

Selected References

Bacon, C.G.; South, D.B. 1989. Chemicals for control of common insect and mite pests in southern pine nurseries. Southern Journal of Applied Forestry. 13: 112–116.

Dixon, W.N.; Barnard, E.L.; Fatzinger, C.W.; Miller, T. 1991. Insect and disease management. In: Duryea, M.L.; Dougherty, P.M., eds., Forest Regeneration Manual. Dordrecht, The Netherlands: Kluwer Academic Publishers: 350–390.

Meeker, J.R. 1997. Recent success and distress over some common arthropod pests in Florida pine nurseries. In: James, R.L., ed. Proceedings of the third meeting of IUFRO Working Party S7.03-4, Diseases and insects in forest nurseries. Report 97-4. Missoula, MT: USDA Forest Service, Northern Region, Forest Health Protection: 3–10.

Rush, P.A.; Hoffard, W.H. 1989. White grubs. In: Cordell, C.E.; Anderson, R.L.; Hoffard, W.H.; Landis, T.D.; Smith, Jr., R.S.; Toko, H.V., tech. coords. Forest nursery pests. Agriculture Handbook 680. Washington, DC: USDA Forest Service: 144–145.

Selman, H.L. 2008. White grubs, *Phyllophaga* and other species. Publication EENY045. Gainesville, FL: University of Florida, Institute of Food and Agricultural Sciences, Entomology and Nematology Department. 4 p.

Speers, C.F.; Schmiedge, D.C. 1961. White grubs in forest tree nurseries and plantations. Forest Pest Leaflet No. 63. Washington, DC: USDA Forest Service. 4 p.

Miscellaneous Pest Problems

Miscellaneous Pest Problems

53. Animal Damage

David B. South

Hosts

Most plant species grown in nurseries can be injured by vertebrates.

Distribution

Seed damage caused by animals occurs worldwide. Damage to plants from browsing animals varies depending on the population present on adjoining lands.

Damage

Seed consumption reduces seed efficiency and profits. Damage to roots is typically minor when compared with seed losses and losses from uprooting by browsers. After the first growing season, clipping by browsers in the winter is often minor and typically has no long-term effects on seedling survival or growth. Browsing late in the season, however, can result in complaints from some customers.

Diagnosis

Birds commonly remove the entire seed from the area. In the absence of tracks, direct observation is often the only evidence that birds are the problem. With large-seeded species, however, empty husks are often a sign of seed predation either by birds or rodents (fig. 53.1). Most managers can identify the animal type by their tracks (fig. 53.2) and fecal pellets (when present). Clipping injury caused by rabbits, hares, and small rodents can usually be identified by a smooth, oblique cut on the woody stem or on the cotyledons, needles, or leaves. Deer browsing injury is characterized by a splintered, ragged break on the stem. Meadow voles remove basal bark from young seedlings, giving them a fuzzy, finely gnawed appearance. The presence of burrowing animals, such as gophers and moles, is detected by mounds that outline tunnels or burrow openings.

Figure 53.1—*A reduction in seed efficiency caused by crows eating pecans during the germination phase.* Photo by David B. South, Auburn University.

Control

Habitat Management

Eliminating protective cover (both within the nursery and in adjacent fields) will reduce favorable habitat for most rodents and other small mammals. Small

Miscellaneous Pest Problems

53. Animal Damage

Figure 53.2—*Turkey tracks.* Photo by David B. South, Auburn University.

mammals are usually wary of crossing long, exposed distances from their burrows to newly established seedbeds. Remove brush, keep grass and weeds short, and eliminate brush piles and trash dumps to reduce protective cover.

Predators

Although rodents, rabbits, and birds are prey for some carnivores, predators cannot keep their populations low enough to prevent damage in nurseries. Predators should be encouraged where their presence does not pose a problem for workers because they influence the behavior of small animals and birds.

Mechanical

Screens or netting can be used to protect seeds in seedbeds and containers (fig. 53.3). A fairly rigid material with fine mesh will keep birds from becoming entangled. If rodents or rabbits are a problem, the netting edge should be buried to prevent digging beneath the barrier. A wire mesh fence 76 to 90 cm (30 to 36 in) tall may help exclude meadow voles and rabbits. Holes should be no larger than 6 mm (0.25 in) for voles and 25 mm (1 in) for rabbits. The bottom 15 cm (6 in) should be turned outward and buried in the ground 15 cm (6 in) deep. The fence must include tight-fitting gates and sills to keep animals from digging below the bottom rails. Some managers use a tall woven wire fence to exclude deer and other large mammals (fig. 53.4). Tight attachment to the ground is important because deer frequently try to go under fences. Electric fences can also be used to deter large browsers.

Trapping is practical for controlling animal pests whose numbers are not too great. Select a trap designed for the animal. It is usually best to bait traps and leave them unset for several days. When the animal is readily taking the bait, set the trap. Trap placement is often extremely important. Tunnels, burrow openings, or holes in a fence are excellent choices for trap placement. Often these trap sets do not require baiting.

Shooting

Shooting can be effective for certain animals whose populations are fairly low. Shooting has the additional advantage of producing frightening sounds. In the past, some nurseries employed bird patrols that warned off birds with shotguns and firecrackers. Bird patrols are a labor-intensive practice, however, which can become very expensive. Before a shooting or trapping program is undertaken, check local laws and regulations governing nuisance animals.

Figure 53.3—*Shade cloth is sometimes used to protect seed from birds.* Photo by David B. South, Auburn University.

Miscellaneous Pest Problems

53. Animal Damage

Figure 53.4—*A high fence can reduce the amount of damage from deer.* Photo by David B. South, Auburn University.

Scare Methods

Firecrackers, propane guns, whistles, horns, scarecrows, flashing lights, and various other devices have been used to frighten birds and deer. However, some animals quickly adapt and eventually ignore these methods. Relocating the devices periodically and combining them with human activity may help them remain effective longer.

Chemical

Repellents can be effective against some birds or animals, but no single chemical is a universal repellent. Area repellents are sprayed on vegetation or cloth strips and placed around a crop to keep animals away. Taste repellents are sprayed directly on a crop to deter feeding. Since new shoot growth is not treated and most chemicals wash off, a single treatment to foliage is usually ineffective. Repellents are often used on seed to deter predation by birds. Repellent use does not require a permit but the product must be registered for such use in the State.

In the past, baits laced with lethal avicides were used to kill birds. Managers now prefer to use sublethal dosages to train crows and other birds to avoid nursery fields. Nursery managers should check with their local county extension agent, U.S. Department of Agriculture, Animal Plant Health Inspection Service (USDA-APHIS), or U.S. Department of the Interior, U.S. Fish and Wildlife Service to determine if a permit is needed before using any product or method to control protected bird species.

Selected References

Conover, M.R. 1989. Relationships between characteristics of nurseries and deer browsing. Wildlife Society Bulletin. 17: 414–418.

Hygnstrom, S.E.; Timm, R.M.; Larson, G.E., eds. 1994. Prevention and control of wildlife damage. Lincoln, NE: University of Nebraska-Lincoln. 2 vols.

Lawrence, W.H.; Kverno, N.B.; Hartwell, H.D. 1961. Guide to wildlife feeding injuries on conifers in the Pacific Northwest. Portland, OR: Western Forestry and Conservation Association. 44 p.

Nielsen, D.G.; Dunlap, M.J.; Miller, K.O. 1982. Pre-rut rubbing by white-tailed bucks: nursery damage, social role and management options. Wildlife Society Bulletin. 10: 341–348.

Radwan, M.A. 1970. Destruction of conifer seed and methods of protection. In: Dana, R.H., ed. Proceedings, Fourth Vertebrate Pest Conference, March 3–5, 1970, Sacramento, CA [Publisher unknown]: 77–82.

Salmon, T.P.; Passof, P.C. 1989. Animal damage. In: Cordell, C.E.; Anderson, R.L.; Hoffard, W.H.; Landis, T.D.; Smith, Jr., R.S.; Toko, H.V., tech. coords. Forest nursery pests. Agriculture Handbook 680. Washington, DC: USDA Forest Service: 152–154.

Scott, J.D.; Townsend, T.W. 1985. Characteristics of deer damage to commercial tree industries of Ohio. Wildlife Society Bulletin. 13: 135–143.

Scott, J.D.; Townsend, T.W. 1985. Deer damage and damage control in Ohio's nurseries, orchards and Christmas tree plantings: the grower's view. In: Bromley, P.T., ed. Proceedings of the Second Eastern Wildlife Damage Control Conference 2: 205–214.

Trent, A.; Nolte, D.; Wagner, K. 2001. Comparison of commercial deer repellents. Tech Tip 0124-2331-MTDC. Missoula, MT: USDA Forest Service, Technology & Development Program, Missoula Technology and Development Center. 6 p.

Miscellaneous Pest Problems

54. Environmental and Mechanical Damage

Michelle M. Cram

Hosts

Abiotic damage from temperature extremes, wind, drought, and mechanical damage can affect all forest nursery seedlings. Seedlings grown from seed sources of a lower latitude or elevation are more susceptible to environmental damage.

Diagnosis

Diagnosis of abiotic damage is often determined by the damage pattern, tissue damage, and records on weather or cultural treatments. Environmental damage tends to be relatively uniform; often affecting a particular crop or seed source more than others (fig. 54.1). Seedlings affected by freeze damage, heat, or drought may go unnoticed for weeks or even months, until stunting or seedling mortality is obvious. Freeze or heat damage will eventually be visible as discolored foliage or inner tissue, stem constrictions, and lesions. Early visual assessment of lethal cold damage is possible following a freeze or unusual winter warming event. Place affected seedlings indoors at room temperature in a plastic bag or other containment to prevent desiccation; after 2 to 8 days look for brown or water-soaked tissue of the stem, buds, needles, or roots (container seedlings). Other more quantitative testing for injury is possible with measuring the level of electrolyte leakage or chlorophyll fluorescence emissions; however, these tests have some drawbacks aside from equipment and technical requirements. Electrolyte leakage must be compared with standard response curves for the species and seed source to be accurate. Chlorophyll fluorescence testing requires green tissue and cannot be used with hardwoods. These tests may not be as accurate in determining lethal damage as the more time consuming visual assessment.

Figure 54.1—*Germinating white oak seedlings were the only species damaged by a December freeze.* Photo by Michelle M. Cram, USDA Forest Service.

Seedling damage related to cultural or mechanical injury usually produces a systematic pattern within the nursery beds. Poor irrigation and shallow undercutting are examples of cultural or mechanical damage that produce systematic patterns. Diagnosis of mechanical damage can usually be confirmed visually. Occasionally, stunting related to irrigation or heat damage can be mistaken as a disease and may require a pathologist or soil testing to rule out pathogens or nematodes.

Specific Problems

Frost

Freeze injury occurs when seedlings are not cold-hardy enough to tolerate freezing temperatures. Foliage damaged by frost will turn from light yellow to red (fig. 54.2). Freeze damaged stem and root tissue becomes discolored and eventually turns the

Figure 54.2—*Eastern white pine foliage damaged by frost.* Photo by Michelle M. Cram, USDA Forest Service.

bark brown or red. A stem constriction can form on a frost-damaged seedling that is able to continue photosynthesis (fig. 54.3). These seedlings can go unnoticed until they become stunted and discolored during the growing season.

Trees become acclimated to withstand freezing temperatures in response to shortened days, lower temperatures (accumulated chill hours), and reduced moisture. Frost damage is more likely in

Miscellaneous Pest Problems

54. Environmental and Mechanical Damage

Figure 54.3—*Freeze injury causing a constriction and inner tissue discoloration of a Fraser fir seedling.* Photo by Michelle M. Cram, USDA Forest Service.

years with unusually warm fall weather that encourages active growth. Similarly, unusual warm periods during the late winter and spring can cause some species or seed sources to break dormancy early, leading to tissue damage when freezing temperatures return.

Frost Heaving

In the fall and spring, soil in nursery beds can freeze during the night and thaw during the day causing the soil and seedlings to lift and fall in response (fig. 54.4). Frost heaving damages seedlings mechanically by breaking the roots and lifting the seedlings out of the ground. First year seedlings and transplants in wet, fine textured soils are most vulnerable to frost heaving.

Winter Burn

Seedling winter burn occurs when seedlings transpire in response to warm, windy weather when the ground is still frozen or dry. Under these conditions seedlings are unable to take up water and the exposed needles and stems become desiccated. Seedling needles turn yellow to red and appear scorched in response to desiccation. Seedlings in containers and in exposed locations have a greater winter burn probability.

Figure 54.4—*Seedlings and soil displaced by frost heaving.* Photo from USDA Forest Service Archives.

Heat Lesions

Seedlings develop lesions or are girdled when ground temperatures reach greater than 52 to 54 °C (126 to 129 °F). Heat lesions on succulent seedlings can range from superficial white spots on stems facing the sunlight to a full constriction at the base (fig. 54.5). Young succulent seedlings with severe heat lesions will collapse and may be confused with damping-off. Older seedlings damaged by heat will often remain erect with a constricted base (fig. 54.6). Damaged seedlings with functioning xylem can become stunted and develop a slight swelling above the heat lesion where carbohydrates accumulate.

Drought

Moisture stress can occur in forest nurseries if irrigation patterns are poor or fail altogether due to either mechanical or human error. Similar damage occurs if seedling transpiration rates exceed the ability of the roots to absorb moisture

Figure 54.5—*Heat lesions on loblolly pine seedlings.* Photo by Michelle M. Cram, USDA Forest Service.

Miscellaneous Pest Problems

54. Environmental and Mechanical Damage

under periods of high heat and dry winds. Slight moisture stress damage can go unnoticed initially. Seedlings that receive less water due to poor irrigation patterns can become stunted, especially in dry years, giving the nursery beds a systematic wavy pattern (fig. 54.7). Other visible symptoms of moisture stress include wilting, graying of foliage, needle and leaf scorch, and premature foliage drop (fig. 54.8). Aboveground symptoms of drought can appear similar to root diseases, injury, or flooding damage.

Wind Abrasion

Seedlings develop lesions and calluses at the groundline in response to soil particles hitting the stems during high winds. Wind abrasion occurs more often in nurseries with sandy soil and in open and exposed fields.

Mechanical

Mechanical injury can be diagnosed based on the damage pattern, cultural records, and interviews with nursery personnel. Damage to seedlings can include severe root pruning, hand weeding, rough handling, and stripped roots at lifting. Seedlings with severe root loss from mechanical injury will have symptoms similar to drought stress.

Prevention

Frost

Freeze injury can often be controlled by sowing species or seed sources that are well adapted to the local conditions. Avoid late summer applications of nitrogen fertilizer and allow seedlings to harden prior to damaging frosts. Protect container seedlings until they are able to

Figure 54.6—*Pine stem girdled by heat and blocking movement of carbohydrates causing the stem to swell above the damaged phloem.* Photo by Edward L. Barnard, Florida Division of Forestry.

Figure 54.7—*Systematic spots of stunted pine seedling caused by a poor irrigation pattern.* Photo by Michelle M. Cram, USDA Forest Service.

Figure 54.8—*Drought damage due to lack of irrigation for 4 days.* Photo by Michelle M. Cram, USDA Forest Service.

Miscellaneous Pest Problems

54. Environmental and Mechanical Damage

withstand freezing cold by maintaining them under cover or insulated from the cold until frost damage is unlikely. Irrigation can be used to prevent frost when the temperatures are above -7 °C (19 °F); however, it is important to apply enough water to continue the freezing process (liquid to solid), which releases heat and maintains seedlings temperature just above freezing.

Frost Heaving

The frequency and severity of frost heaving can be reduced with cultural practices that increase soil drainage and prevent rapid soil temperature fluctuation between freeze and thaw. Mulch use insulates the ground and reduces the effect of the day-to-night temperature changes. Shade in the form of manmade covers or natural shading by vegetation will also protect seedlings from frost heaving. Uniform seedling beds with larger root systems and full crown closure will provide natural shading and reduce frost heaving.

Winter Burn

Windbreaks help protect overwintering seedlings from drying winds that lead to winter burn. Fencing or vegetation windbreaks work best when perpendicular to the direction of the wind. Other cultural techniques to protect seedlings from winter burn include thick mulches, shade cloth, bed frames, or cold protection fabric. In North Central States, snow blowers have been used to coat seedlings in snow to protect from winter burns (fig. 54.9). In Southern States, overwintering conifers may be subjected to unusual warm periods (18 to 28 °C, 64 to 82 °F) and require irrigation if beds become too dry.

Heat Lesions

Seedbed orientation along the sun's summer path, plus optimum seedling density allows for mutual shading. Sow seedlings early enough so that seedlings are older when temperatures reach damaging levels and have a protective bark layer. If a crop must be sown late in the growing season, shading may be required for young and vulnerable seedlings. Frequent irrigation during extreme heat periods can be used to cool ground temperatures. Avoid using dark-colored mulches that absorb sunlight.

Drought

Seedling damage from drought can be avoided through proper irrigation. Uniform water distribution is especially important to avoid under-irrigated areas. Maintain soil water potential to field capacity and carefully monitor soil water during seedling germination and early growth. Increasing the soil organic matter of well-drained soils will increase the soil's water holding capacity. Mulch use on seedbeds can help prevent evaporation and reduce water runoff.

Wind Abrasion

Windbreaks reduce soil particle movement that causes wind abrasion damage. Soil stabilizers such as mulches and polymeric adhesives can significantly reduce soil movement. Other cultural techniques to stabilize soil include irrigating the soil surface and sowing cover crops in fields out of production.

Figure 54.9—*Snow blowers are used in Northern States to coat seedlings in snow for protection from winter burns.* Photo from Minnesota Department of Natural Resources.

Mechanical

Well-trained and skilled nursery equipment operators are essential to avoiding mechanical damage. The condition of the soil, seedlings, and weather should be taken into consideration prior to any cultural or lifting operations. Ensure that seedlings have adequate field moisture following mechanical treatments to the root systems to prevent moisture stress.

Selected References

Anekonda, T.S.; Adams, W.T. 2000. Cold hardiness testing for Douglas-fir tree improvement programs: guidelines for a simple, robust, and inexpensive screening method. Western Journal of Applied Forestry. 15: 129–136.

Banard, E.L. 1990. Groundline heat lesions on tree seedlings. Florida Department of Agriculture & Consumer Services, Division of Plant Industry. Plant Pathology Circular 338. 2 p.

Carlson, W.C.; Anthony, J.G.; Plyler, R.P. 1987. Polymeric nursery bed stabilization to reduce seed losses in forest nurseries. Southern Journal of Applied Forestry. 11: 116–119.

Goulet, F. 1995. Frost heaving of forest tree seedlings: a review. New Forests. 9: 67–94.

Miscellaneous Pest Problems

54. Environmental and Mechanical Damage

Hermann, R.K. 1990. Cold injury: frost damage, frost heaving, winter desiccation. In: Hamm, P.B.; Campbell, S.J.; Hansen, E.M., eds. Growing healthy seedlings: identification and management of pests in northwest forest nurseries. R6-PNW-019-90. Corvallis, OR: Oregon State University, Forest Research Laboratory and USDA Forest Service, Pacific Northwest Region: 68–72.

Landis, T.D. 1989. Environmental and mechanical damage. In: Cordell, C.E.; Anderson, R.A.; Hoffard, W.H.; Landis, T.D.; Smith, Jr., R.S.; Toko, H.V., tech. coords. Forest nursery pests. Agriculture Handbook 680. Washington, DC: USDA Forest Service: 155–158.

McKay, H.M. 1996. A review of the effect of stresses between lifting and planting on nursery stock quality and performance. New Forests. 13: 363–393.

Smit-Spinks, B.; Swanson, B.T.; Markhart, III, A.H. 1984. The effect of photoperiod and thermoperiod on cold acclimation and growth of *Pinus sylvestris*. Canadian Journal of Forest Research. 15: 453–460.

Timmis, R.; Flewelling, J.; Talbert, C. 1994. Frost injury prediction model for Douglas-fir seedlings in the Pacific Northwest. Tree Physiology. 14: 855–869.

Miscellaneous Pest Problems

55. Pesticide Injury
David B. South

Hosts

Pesticide injury can occur on all tree species.

Damage

Injury can range from slight cosmetic effects to mortality.

Diagnosis

By maintaining untreated check plots, the nursery manager can determine if seedling injury is due to pesticides (fig. 55.1). Without check plots, diagnosis can be difficult, even for a specialist. Successful diagnosis without check plots requires a thorough knowledge of the employed pesticide's modes of action and experience with injury from inert ingredients, insects, and certain abiotic factors.

Figure 55.1—*Pesticide injury can be easily detected by use of untreated check plots.* Photo by David B. South, Auburn University.

Injury Types

Seedling injury varies with the chemical used; the adjuvant used; inert ingredients; the concentration contacting the seedling; the species, age, and physiological condition of the seedling; the soil texture; the weather; and other environmental considerations. Pesticide damage is most likely to occur when seedling tissue is young and succulent.

Seed

Some chemical seed treatments may inhibit germination. In some nursery trials, the reduction caused by a fungicide seed treatment may be economically significant (for example, 7-percent lower germination) but not statistically significant.

Foliage

Newly formed needles may be burned by inert ingredients like naphtha (fig. 55.2). In some situations, nitrogen fertilization can increase succulence of new tissue, thus increasing susceptibility to injury from certain herbicides. Some managers have observed injury to foliage after fumigating soil with compounds that release methyl isocyanate (MITC).

Roots

Certain herbicides can damage seedling roots. Several dinitroaniline herbicides can injure root systems if the herbicide is incorporated into the soil before sowing. Certain herbicides (for example, sulfurometuron) can also affect the root system when applied to the soil surface. When applied directly to roots after lifting, some fungicides may reduce root growth after transplanting.

Stems

When applied to the soil surface, some dinitroaniline herbicides may cause herbicide galls to form on the stem at the root collar (fig. 55.3). Occasionally, some fungicides (for exaple, dicloran) can also injure stems when applied to newly emerged seedlings.

Growth

Some pesticide injury will be mostly cosmetic (fig. 55.1) and may have little effect on growth. Cosmetic injury may be typical with some pesticides. In some cases, however, growth has been stunted when seedlings are growing in soil containing an herbicide that has a long half-life.

Miscellaneous Pest Problems

55. Pesticide Injury

Figure 55.2—*Seedling injury can be caused by ingredients that are incorrectly labeled as "inert." All seedlings in this photo were treated with an insecticide containing naphtha, but only the young seedlings were injured.*
Photo by David B. South, Auburn University.

Figure 55.3—*Herbicide gall on loblolly pine seedling.*
Photo by David B. South, Auburn University.

Reason for Injury

Damage caused by pesticides usually results from one of the following situations.

Misapplication

Injury to a seedling crop is often the result of failure to carefully read the pesticide label. For example, one manager injured pine seedlings when applying a wettable dispersible granule as though it was a granular herbicide. Additional mistakes include failure to properly clean equipment, inadequate herbicide solution mixing, improper equipment calibration, improper application timing, and failure to seek expert advice.

Particle Drift

Particle drift involves airborne droplet movement away from the intended target. The amount of drift is influenced by droplet size, microclimate, chemical formulation, and adjuvants. Nozzle type and size selection is critical. In general, the smaller the spray droplets are, the greater the drift. Hence, when applying pesticides like glyphosate, a coarse spray (large droplets) should be chosen. Thickeners, additives, foaming agents, and emulsifying agents can be added to the solution to affect droplet size and lessen drift.

Vapor Drift

Vapor drift is applied as a solid or liquid to the target site and then a portion moves offsite in a gaseous state. In some cases, this transformation can occur days after the initial application. As a result, some managers do not use certain herbicides (for example, oxadiazon and oxyfluorfen) in greenhouses since they might "lift off" and injure susceptible plants. Also, weather conditions can profoundly influence offsite drift. For example, an inversion layer can suspend and move MITC vapor a considerable distance before contacting sensitive seedlings. To reduce this injury, some experts recommend covering certain soil fumigants (for example, metam-sodium) with a tarp.

Incorrect Formulation

Careful selection of the pesticide formulation can minimize plant damage when the product is applied directly to seedlings. Pesticides formulated as granules or water-dispersible packets are less likely to cause damage than emulsifiable concentrates. Pesticide formulations and adjuvants containing oil can also injure succulent foliage.

Adverse Weather

Temperature, rainfall, inversion layers, and cloud cover can affect pesticide injury. Some pesticides cause damage if applied during hot or dry weather. During heavy rainfall events, some water-soluble herbicides may be carried in surface

Miscellaneous Pest Problems

55. Pesticide Injury

runoff and may accumulate in low spots. With certain herbicides, injury may result after high winds that sandblast seedlings (fig. 55.1), especially if the seedlings are less than 2 months old.

Research and education are keys to preventing pesticide injury. Seeking expert advice before applying a pesticide will likely avoid costly injury. When using selective pesticides, well-informed managers can often lower the pest population and minimize the risk of seedling injury.

Selected References

Buzzo, R.J. 2003. Phytotoxicity with metam sodium. In: Riley, L.E.; Dumroese, R.K.; Landis, T.D., tech. coords. National proceedings: forest and conservation nursery associations—2002. Proceedings RMRS-P-28. Ogden, UT: USDA Forest Service, Rocky Mountain Research Station: 79–83.

Callan, N.W. 1979. Dacthal injury on Douglas-fir and true firs at the Medford Forest Nursery. In: Jones, Jr., E.P., ed. Proceedings, 2nd Biennial Southern Silvicultural Research Conference, November 4–5, 1982, Atlanta, GA. GTR-SE-24. Asheville, NC: USDA Forest Service, Southeastern Forest Experiment Station: 418–426.

Fisher, J.W.; Landis, T.D. 1990. Dicloran fungicide causes stem injury to container spruce seedlings. Tree Planters' Notes. 40(1): 39–42.

Landis, T.D. Pesticide phytotoxicity. Chapter 30. In: Hamm, P.B.; Campbell, S.J; Hansen, E.M., eds. Proceedings. Growing healthy seedlings. PNW-019-90. Corvallis, OR: USDA Forest Service, Pacific Northwest Region: 79–83.

Linderman, R.G.; Davis, E.A.; Masters, C.J. 2008. Efficacy of chemical and biological agents to suppress *Fusarium* and *Pythium* damping-off of container-grown Douglas-fir seedlings. Plant Health Progress. doi:10.1094/PHP-2008-0317-02-RS.

Rowan, S.J. 1978. Treflan injury of loblolly pine seedlings. Tree Planters' Notes. 29(3): 25–26.

Samtani, J.B.; Masiunas, J.B.; Applyby, J.E. 2008. Injury on white oak seedlings from herbicide exposure simulating drift. Horticulture Science. 43(7): 2076–2080.

South, D.B. 1980. Nurserymen must leave herbicide check plots. In: Proceedings, 1980 Southern Nursery Conference, September 2–4, 1980, Lake Barkley, KY. Atlanta, GA: USDA Forest Service: 123.

South, D.B.; Hill, T. 2009. Results from six *Pinus taeda* nursery trials with the herbicide pendimethalin in the USA. Southern Forests. 71(3): 179–185.

South, D.B.; Kelley, W.D. 1982. Effects of selected pesticides on short-root development of greenhouse-grown *Pinus taeda* seedlings. Canadian Journal of Forest Research. 12: 29–35.

South, D.B.; Mexal, J.G. 1983. Effect of bifenox and oxyfluorfen on emergence and mortality of loblolly seedlings under growth chamber conditions. In: Jones, Jr., E.P.; ed. Proceedings, 2nd Biennial Southern Silvicultural Research Conference, November 4–5, 1982, Atlanta, GA. GTR-SE-24. Asheville, NC: USDA Forest Service, Southeastern Forest Experiment Station: 418–426.

Taylor, Jr., J.W.; South, D.B. 1989. Pesticide injury. In: Cordell, C.E.; Anderson, R.L.; Hoffard, W.H.; Landis, T.D.; Smith, Jr., R.S.; Toko, H.V., tech. coords. Forest nursery pests. Agriculture Handbook 680. Washington, DC: USDA Forest Service: 159–161.

Miscellaneous Pest Problems

56. Salinity Damage
Katy M. Mallams and Thomas D. Landis

Hosts

Many native plants, including most forest tree species, are very sensitive to the level of salts in soil and irrigation water. Germinants and young seedlings are particularly sensitive due to their succulent nature and small root systems. Older seedlings and transplants become more tolerant of salinity. Small-seeded or slow-growing species are more susceptible to salt damage because they take longer to grow out of the juvenile succulent stage.

Distribution

Soluble salts are inorganic chemical compounds that release electrically charged particles called ions when they are dissolved in water. The specific compounds present and total salt concentration in soil or water will determine whether soluble salts have a beneficial, neutral, or damaging effect on seedlings. For example, magnesium sulfate ($MgSO_4$) can be used as a fertilizer, but sodium chloride (NaCl) is toxic.

Soluble salts in nurseries can originate from several sources. Salts accumulate naturally in the surface soil of arid and semiarid areas with less than 38 cm (15 in) of annual precipitation and in areas where evapotranspiration exceeds precipitation. Groundwater used for irrigation in these areas is often high in salts. In coastal areas, saltwater may intrude into the groundwater. Overfertilization, improper use of fertilizers, soil compaction, naturally calcareous soil, and contaminated mulches can result in excessive salt levels in nursery soils and container media.

Damage

There are four ways that soluble salts can injure seedlings. High concentrations of soluble salts can reduce the amount of water available to seedlings through osmosis, the process that enables plants to absorb water from the soil. High concentrations of sodium ions reduce the permeability and water infiltration rate of the soil. High concentrations of sodium, chloride, and boron are directly toxic to plants. An imbalance of nutrients, such as calcium, will reduce the availability of other essential nutrients such as iron, magnesium, and phosphorus.

Salinity damage can result in mortality of young germinants and growth losses in older stock. Growth losses may be difficult to determine because significant reductions in growth often occur before more visible symptoms become evident. Seedlings that are stunted due to salinity damage may perform poorly after outplanting due to low vigor or poorly developed root systems. High concentrations of salts may also reduce frost hardiness and resistance to some pathogenic fungi. In addition to damaging plants, high concentrations of soluble salts can form deposits that clog sprinkler nozzles and leave unsightly whitish deposits on foliage.

Hot, dry weather increases the potential for salinity damage. Irrigation practices that enable the soil to dry out concentrate salts in the soil. In particular, brief irrigation used for cooling seedlings in hot weather increases salinity at the soil's surface due to evaporation, which pulls salts to the surface where they concentrate. Soil management practices can also lead to accumulation of salts in the seedling root zone, especially in fine-textured soils. Improper or excessive cultivation may break down the soil structure, reducing porosity, inhibiting water infiltration and drainage, and allowing salts to accumulate. Repeated use of heavy equipment in seedbeds causes impermeable hard pans, which can restrict drainage. Application of dry inorganic fertilizers without sufficient irrigation can concentrate salts at the soil surface where the roots of young germinants are present.

Diagnosis

Mortality due to salt damage in germinating seeds and emerging seedlings is often confused with other diseases such as damping-off. The pattern of mortality is often patchy and irregular. In larger seedlings, symptoms vary depending on the species, stock type, and age. Symptoms include chlorosis (fig. 56.1), scorched needle tips and leaf margins (fig. 56.2), bluish leaf color, white

Figure 56.1—*Chlorosis of young needles, a symptom of iron deficiency associated with saline soil.*
Photo by Thomas D. Landis, USDA Forest Service.

Miscellaneous Pest Problems

56. Salinity Damage

Figure 56.2—*Scorched needle tips caused by physiological drought due to high levels of soluble salts.* Photo by Thomas D. Landis, USDA Forest Service.

deposits on foliage (fig. 56.3), stunting (fig. 56.4), lack of roots near the surface, patchy growth patterns (fig. 56.5), and mortality. Salt crystals can sometimes be seen on seedling roots when examined under a microscope. Soil characteristics may also provide evidence for salinity damage. Calcareous soils often develop a white crust on the surface (fig. 56.6) that effervesces when tested with a drop of dilute acid. Soil high in sodium salts is blackish with a slick feel.

In most cases, salinity damage is very difficult to identify solely by visual symptoms. Accurate diagnosis requires comprehensive evaluation that should include chemical analysis of the soil or container media, irrigation water, and seedling foliage. Soil samples should be collected at several different depths, but especially from the surface horizons where salts accumulate. An electrical conductivity test of the irrigation water or a saturated paste of the soil is the best way to measure the total soluble salt level. Electrical conductivity can be determined at the nursery using a conductivity meter. Readings of greater than 4,000 microSiemens per centimeter (mS/cm) are considered excessive, and values between 2,500 and 4,000 mS/cm should be of concern (fig. 56.7). Concentrations of individual salts

Figure 56.3—*Salt accumulation at the tips of spruce needles.* Photo by Thomas D. Landis, USDA Forest Service.

can be measured by laboratory tests of water extracted from the soil. Salinity can be described in terms of total salinity, or concentrations of specific salt ions. High pH may also be an indicator of excessive salt levels.

Control

The best solution for problems with salinity is to avoid them through careful selection of nursery sites. During the site selection process, both the irrigation water and soil should be tested for total soluble salts and the relative concentrations of specific salts. Soil permeability and porosity should also be tested to determine the leaching potential of the soil. In established nurseries, the soil should be mapped to identify problem areas and ensure that sensitive species and stock types are planted on the best soils. Irrigation water should be tested several times a year to monitor changes in salinity.

Figure 56.4—*Stunted seedlings, often in a variable pattern, are a characteristic symptom of saline soil.* Photo by Thomas D. Landis, USDA Forest Service.

Miscellaneous Pest Problems

56. Salinity Damage

Figure 56.5—*Stunting and chlorosis often occur in irregular patches throughout the seedbed.* Photo by Thomas D. Landis, USDA Forest Service.

Figure 56.6—*Salt crust on the surface is an indicator of saline soil.* Photo by Thomas D. Landis, USDA Forest Service.

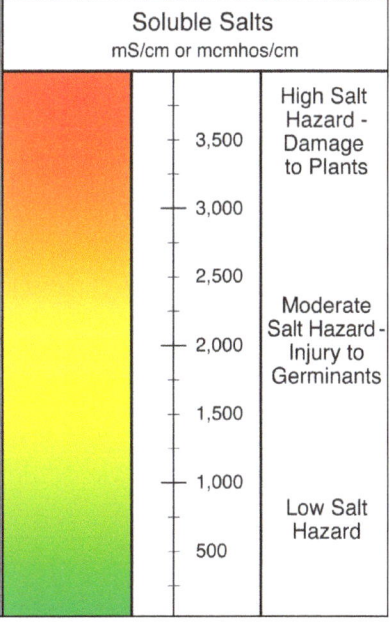

Figure 56.7—*Guidelines for interpreting soluble salt concentrations.* Graphic by Thomas D. Landis, USDA Forest Service.

Cultural

In bareroot nurseries, cultural practices such as deep-ripping, addition of organic matter, and leaching of salts can alleviate the effects of saline water and prevent buildup of salts in the soil. Leaching consists of applying large amounts of water to the soil to dissolve salts and flush them down below the seedling root zone. This treatment should be conducted while fields are fallow and after deep-ripping to break up any impermeable hard pans and promote rapid drainage.

Deep irrigation to decrease the levels of salts in the surface soil during germination and seedling emergence can reduce the potential for damage. It is important, however, to ensure that the soil does not become waterlogged during this treatment or seedlings may become vulnerable to root diseases. It is always best to irrigate during times when the evapotranspiration rate is low to avoid salt buildup in the surface soil.

Mulches are a good way to maintain consistent soil moisture, decrease evaporation from the soil, improve water infiltration, and prevent the formation of salt crusts. Light-colored mulches are recommended to avoid heat damage to seedlings that may occur with the use of dark-colored mulches. A thin layer of fresh sawdust or hydromulch, about 1 cm thick, may be used as mulch on top of newly sown seedbeds or on seedlings after emergence. Sand or grit can also be effective as a mulch but should be tested for salts before use.

Inorganic fertilizers are salts, so it is critical that they be applied carefully because they add to the total salt level. In nurseries

Forest Nursery Pests

Miscellaneous Pest Problems

56. Salinity Damage

with naturally saline soil or water, only products with a low salt index should be used. In container nurseries, use of very dilute liquid fertilizer or controlled-release fertilizer is recommended to keep salts within acceptable limits. Several small fertilizer applications are less likely to cause damage than one large presowing application. Organic fertilizers such as sewage sludge and animal manure should be tested before they are purchased as they may contain high levels of salts, heavy metals, or even poisons.

Chemical

Chemical treatments should be applied only after thorough testing to avoid creating additional problems. Soils with excessive levels of sodium can be treated by adding gypsum at a rate of about 22.4 metric tons per ha (10 tons per acre). Adding gypsum will improve the porosity and infiltration rate of the soil. Soils high in calcium can be treated with elemental sulfur at a rate of 560 to 1,120 kgs per ha (500 to 1,000 lbs per acre). Elemental sulphur converts calcium carbonate to the more soluble calcium sulfate. Sulfur treatments also lower soil pH but, because the process takes time, it should be applied at least 1 year before a crop is sown. A significant lowering of soil pH may take many years to accomplish.

Selected References

Hamm, P.B.; Campbell, S.J.; Hansen, E.M. 1990. Growing healthy seedlings: identifications and management of pests in northwest forest nurseries. Special Publication 19. Corvallis, OR: Oregon State University, Forest Research Laboratory. 110 p.

Landis, T.D. 1989. Salinity damage. In: Cordell, C.E.; Anderson, R.A.; Hoffard, W.H.; Landis, T.D.; Smith, Jr., R.S.; Toko, H.V., tech. coords. Forest nursery pests. Agriculture Handbook 680. Washington, DC: USDA Forest Service: 150–151.

Landis, T.D.; Tinus, T.W.; McDonald, S.E.; Barnett, J.P. 1994. The container tree nursery manual, volume four: seedling nutrition and irrigation. Agriculture Handbook 674. Washington, DC: USDA Forest Service. 119 p.

Sinclair, W.A.; Lyon, H.H.; Johnson, W.T. 1987. Diseases of trees and shrubs. Ithaca, NY: Cornell University Press. 575 p.

Directory of Authors and Coordinators

Art Antonelli	Extension Entomologist (retired), Professor Emeritus, Washington State University Research and Extension Center, 2606 West Pioneer, Puyallup, WA 98371
Edward L. Barnard	Forest Pathologist (retired), Florida Department of Agriculture and Consumer Services, 503 NW 123rd Street, Newberry, FL 32669
Sonja Beavers	National Publications Editor, USDA Forest Service Washington Office, Sidney R. Yates Federal Building, 201 14th Street SW, Washington, DC 20024
Gary Chastagner	Plant Pathologist, Washington State University Research and Extension Center, 2606 West Pioneer, Puyallup, WA 98371
Michelle M. Cram	Plant Pathologist, USDA Forest Service Southern Region, Forest Health Protection, 320 Green Street, Athens, GA 30602
Judith F. Danielson	Genetics Forester (retired), USDA Forest Service, Dorena Genetic Resource Center, 34963 Shoreview Drive, Cottage Grove, OR 97424
Wayne N. Dixon	Assistant Director, Division of Plant Industry, Florida Department of Agriculture and Consumer Services, P.O. Box 147100, Gainesville, FL 32614
Coleman Doggett	Entomologist (retired), State of North Carolina and Duke University, 217 Rosecommon Lane, Cary, NC 27511
R. Kasten Dumroese	Research Plant Physiologist and National Nursery Specialist, USDA Forest Service Rocky Mountain Research Station, 1221 South Main Street, Moscow, ID 83843
Marianne Elliott	Research Associate, Washington State University Research and Extension Center, 2606 West Pioneer, Puyallup, WA 98371
Scott A. Enebak	Professor of Forest Pathology and Director, Southern Forest Nursery Management Cooperative, Auburn University, 3301 Forestry and Wildlife Sciences Building, Auburn, AL 36849
Stephen W. Fraedrich	Research Plant Pathologist, USDA Forest Service Southern Research Station, 320 Green Street, Athens, GA 30605
Michelle S. Frank	Entomologist, USDA Forest Service Northeastern Area, Forest Health Protection, 11 Campus Boulevard, Suite 200, Newtown Square, PA 19073
Tom Gordon	Professor and Chair, Department of Plant Pathology, University of California, One Shields Avenue, Davis, CA 95616
Diane L. Haase	Western Nursery Specialist, USDA Forest Service Pacific Northwest Region, 333 SW First Avenue, Portland, OR 97204
Everett M. Hansen	Professor Emeritus, Forest Pathology, Department of Botany and Plant Pathology, Oregon State University, 2082 Cordley Hall, Corvallis, OR 97331
Charles S. Hodges	Pathologist, Professor Emeritus, North Carolina State University, 2510 Thomas Hall, Raleigh, NC 27695
Robert L. James	Plant Pathologist, Plant Disease Consulting Northwest, 520 SE Columbia River Drive, Suite 116, Vancouver, WA 98661
Jennifer Juzwik	Research Plant Pathologist, USDA Forest Service Northern Research Station, 1561 Lindig Street, St. Paul, MN 55108
Thomas D. Landis	National Forest Nursery Specialist (retired) and consultant, 3248 Sycamore Way, Medford, OR 97504
Will R. Littke	Plant Pathologist, Weyerhaeuser Company, P.O. Box 9777, Federal Way, WA 98063-9777
Katy M. Mallams	Plant Pathologist (retired), USDA Forest Service Pacific Northwest Region, Forest Health Protection, 2606 Old Stage Road, Central Point, OR 97502
Alex C. Mangini	Entomologist, USDA Forest Service Southern Region, Forest Health Protection, Alexandria Field Office, 2500 Shreveport Highway, Pineville, LA 71360

Directory of Authors and Coordinators

Albert E. Mayfield, III	Research Entomologist, USDA Forest Service Southern Research Station, 200 W.T. Weaver Boulevard, Asheville, NC 28804
Kathleen McKeever	Graduate Research Associate, Washington State University Research and Extension Center, 2606 West Pioneer, Puyallup, WA 98371
John T. Nowak	Entomologist, USDA Forest Service Southern Region, Forest Health Protection, 200 W.T. Weaver Boulevard, Asheville, NC 28804
Forrest L. Oliveria	Field Office Representative, USDA Forest Service Southern Region, Forest Health Protection, Alexandria Field Office, 2500 Shreveport Highway, Pineville, LA 71360
Michael E. Ostry	Research Plant Pathologist, USDA Forest Service Northern Research Station, 1992 Folwell Avenue, St. Paul, MN 55108
Jill D. Pokorny	Plant Pathologist, USDA Forest Service Northeastern Area, Forest Health Protection, 1992 Folwell Avenue, St. Paul, MN 55108
Lee E. Riley	Horticulturist, USDA Forest Service Pacific Northwest Region, Dorena Genetic Resource Center, 34963 Shoreview Drive, Cottage Grove, OR 97424
James D. Solomon	Research Entomologist (retired), USDA Forest Service, Center for Bottomland Hardwood Research, 432 Stoneville Road, Stoneville, MS 38776
David B. South	Professor of Forest Regeneration (retired), Auburn University, 3301 Forestry and Wildlife Sciences Building, Auburn, AL 36849
Glen R. Stanosz	Professor of Plant Pathology, Department of Plant Pathology, University of Wisconsin-Madison, 1630 Linden Drive, Madison, WI 53706
Tom Starkey	Research Fellow, Southern Forest Nursery Management Cooperative, Auburn University, 3301 Forestry and Wildlife Sciences Building, Auburn, AL 36849
John W. Taylor, Jr.	Integrated Pest Management Specialist, USDA Forest Service Southern Region, Forest Health Protection, 1720 Peachtree Road NW, Room 816N, Atlanta, GA 30309
Michael Taylor	Assistant Manager, IFA Nurseries, Inc., 1887 North Holly Street, Canby, OR 97013
Jerry E. Weiland	Research Plant Pathologist, USDA Agricultural Research Service, Horticultural Crops Research Laboratory, 3420 NW Orchard Avenue, Corvallis, OR 97330

Index of Nursery Pests

This index lists the pests mentioned in the numbered chapters. It includes scientific names of the pathogens and insects, as well as common names of diseases; insects; other biotic pests such as mammals; and environmental factors, such as frost. The number(s) after the name of the pest refers to the **chapter(s)** where information can be found.

Abiotic damage, 54
Actinomycetes, 34
Amphimallon majalis, 52
Animal damage, 53
   birds, 20, 44, 47, 53
   deer, 53
   gophers, 53
   hares, 53
   meadow voles, 53
   moles, 53
   rabbits, 53
   rodents, 53
Anomala spp., 52
Anoplonyx occidens, 22
Anthracnose, 23, 39
Anurogryllus spp., 44
   A. arboreus, 44
Aphids, 42, 48
Apiognominia spp., 23
Aspergillus spp., 39
Asperosporium sequoiae, 12
Aureobasidium spp., 23
Bacteria, 24, 34, 35, 41
Bacterial leaf spot, 24
Barbara spp., 49
Beetles, 29, 30, 49, 50, 52
Black root rot, 31
Blackstem disease, 26
Blister rust, 19
Borers, 29, 46, 49
Botrytis blight, 35, 40
Botrytis cinerea, 35
Brachyrhinus spp., 51
Brown felt blight, 18
Brown spot needle blight, 1, 4
Calonectria kyotensis, 32
Caloscypha fulgens, 39
Cankers, 3, 10, 11, 13, 14, 16, 19, 23, 24, 26, 29, 34, 35, 39, 40
Ceratobasidium spp., 15
Cercospora blight, 9
Cercospora sequoiae, 9, 12
Cercospora sequoiae var. *juniperi*, 9
Charcoal root rot, IPM, 31
Chlorosis, 28, 32, 34, 36, 37, 38, 41, 56
Chrysomela scripta, 30
Cinara spp., 42

Clearwing borer, 29
Cold temperatures, 41, 54
Collar rot, 3, 10, 31, 37
Colletotrichum spp., 23
Cone beetles, 49
Cone borers, 49
Cone moths, 49
Coneworms, 49
Conophthorus spp., 49
   C. coniperda, 49
Contarinia spp., 49
   C. oregonensis, 49
Cotalpa spp., 52
Cottonwood borer, 29
Cottonwood leaf beetle, 30
Crambidae, 20
Cranberry girdler, 20
Crickets, 44, 47
Cronartium quercuum, 5
Cronartium quercuum f.sp. *fusiforme*, 6
Cronartium ribicola, 19
Cryptocline spp., 23
Crysoteuchia topiaria, 20
Curculio spp., 49
Curculionidae, 51
Cutworms, 43
Cyclaneusma minus, 8
Cydia spp., 49
   C. latiferranea, 49
Cylindrocarpon root disease, 2
Cylindrocarpon spp., 2
   C. cylindroides, 2
   C. destructans, 2
   C. didymum, 2
   C. tenue, 2
Cylindrocladiella spp., 32
Cylindrocladium diseases, 32
Cylindrocladium spp., 32
   C. floridanum, 32
   C. parasiticum, 32
   C. scoparium, 32
Cytospora spp., 26
   C. chrysosperma, 26
Damping-off, 3, 10, 14, 15, 31, 32, 33, 39, 43, 54, 56
Dark-winged fungus gnat, 45
Diaporthe lokoyae, 13

Dichelonyx spp., 52
Dingy cutworm, 43
Dioryctria spp., 49
Diplodia canker, 3
Diplodia collar rot, 3
Diplodia pinea, 3, 39
Diplodia scrobiculata, 3
Diplodia shoot blight, 3, 17
Diplodina spp., 23
Diplotaxis spp., 52
Diprionidae, 22
Discella spp., 23
Discula spp., 23, 39
   D. destructiva, 39
Dothichiza spp., 26
   D. populea, 26
Dothiorella spp., 46
Dothistroma needle blight, 1, 4
Dothistroma pini, 4
Dothistroma septosporum, 4
Douglas-fir cone gall midge, 49
Douglas-fir seed chalcid, 49
Drepanopeziza spp., 25
Drought, 13, 46, 52, 54, 56
Eastern gall rust, 5
Elasmopalpus lignosellus, 46
Endocronartium harknessii, 5
Environmental damage, 41, 54
Eotetranychus spp., 50
Eriophyoid mites, 50
Eriophyoidea, 50
Eruptio acicola, 1
Erysiphe spp., 28
Eucosma spp., 49
Feltia ducens, 43
Field crickets, 44
Filbertworm, 49
Foliage blight, 10, 11, 12, 18, 32, 35, 40
Freeze injury, 54
Frost, 7, 13, 23, 35, 40, 54
Fungus gnats, 14, 45
Fusarium root disease, 34
Fusarium spp., 2, 14, 26, 31, 33, 34, 38, 39
   F. acuminatum, 34
   F. avenaceum, 34
   F. circinatum, 14, 39
   F. commune, 34

Index of Nursery Pests

F. equiseti, 34
F. moniliforme var. *subglutinans*, 14
F. oxysporum, 34, 39
F. proliferatum, 34
F. "roseum" complex, 34
F. sambucinum, 34
F. solani, 26, 34
F. sporotrichioides, 34
F. subglutinans, 14, 39
Fusarium stem disease, 34
Fusiform rust, 5, 6
Gall midges, 49
Gall mites, 50
Gall rust, 5
Galls, 5, 6, 29, 36, 42, 46, 48, 49, 50, 55
Gliocladium spp., 35
Glomerella spp., 23
Gnominia spp., 23
Gnomoniella spp., 23
Gnophothrips fuscus, 49
Gray mold, 35
Gremmeniella abietina, 16
Gryllotalpidae, 47
Gryllus spp., 44
Heat, 26, 41, 54, 56
Herpotrichia juniperi, 18
Hoplolaimus spp., 36
Hylobius pales, 51
Hypocotyl rot (see Fusarium), 34
June beetles, 52
Kabatina juniperi, 12
Larch needle cast, 7
Lasiodiplodia theobromae, 26, 39
Leaf beetle, 30
Leaf blight, 23, 24, 40
Leaf rust, 27
Leaf spots, 23, 24, 25, 26, 32, 40
Leaffooted pine seed bug, 49
Lecanosticta acicola, 4
Lepidoptera spp., 43
Leptoglossus spp., 49
 L. corculus, 49
 L. occidentalis, 49
Lesser cornstalk borer, 46
Longidorus spp., 36
 L. americanus, 36
Lophodermium spp., 8
 L. seditiosum, 8
Lophophacidium hyperboreum, 18
Lower stem canker (see Fusarium), 34
Lygus bugs, 48

Lygus spp., 48
 L. elisus, 48
 L. hesperus, 48
 L. lineolaris, 48
Macrophomina phaseolina, 31
Marssonina blight, 25
Marssonina spp., 25
 M. balsamiferae, 25
 M. brunnea f.sp. *brunnea*, 25
 M. brunnea f.sp. *trepidae*, 25
 M. castagnei, 25
 M. populi, 25
May beetles, 52
Mechanical damage, 54
Megastigmus spermotrophus, 49
Melampsora spp., 27
 M. medusae, 27
 M. occidentalis, 27
Meloidodera spp., 36
Meloidogyne spp., 36
Meria laricis, 7
Microsphaera spp., 28
Miridae, 48
Mole cricket, 47
Moths, 20, 21, 29, 43, 46, 49
Mucor spp., 39
Mycetophilidae, 45
Mycoplasmas, 41
Mycosphaerella dearnessii, 1
Mycosphaerella pini, 4
Nantucket pine tip moth, 21
Needle blight, 1, 4
Needle cast, 1, 7, 8
Nematodes, 31, 36, 41, 54
 dagger nematode, 36
 ectoparasitic nematodes, 36
 endoparasitic nematodes, 36
 lance nematode, 36
 lesion nematode, 36
 needle nematode, 36
 pine cystoid nematode, 36
 root-knot nematode, 36
 stubby-root nematode, 36
 stunt nematode, 36
Nemocastes incomptus, 51
Neodiprion spp., 22
 N. burkei, 22
 N. lecontei, 22
 N. pinetum, 22
 N. sertifer, 22
 N. similis, 22
 N. tsugae, 22

Neopeckia coulteri, 18
Noctuidae, 43
Odontopus calceatus, 51
Oidium spp., 28
Oligonychus spp., 50
 O. bicolor, 50
 O. coniferarum, 50
 O. ilicis, 50
 O. milleri, 50
 O. platani, 50
 O. subnudus, 50
 O. ununguis, 50
Oomycetes, 33, 37, 38, 40
Otiorhynchus spp., 51
 O. ovatus, 51
 O. rugusostriatus, 51
 O. sulcatus, 51
Ovulariopsis spp., 28
Pachylobius picivorous, 51
Paranthrene dollii, 29
Passalora blight, 9
Passalora sequoiae, 9, 12
Penicillium spp., 39
Peridermium harknessii, 5
Pestalotia spp., 10, 39
Pestalotiopsis foliage blight, 10
Pestalotiopsis funerea, 10
Pesticide injury, 55
Phacidium abietis, 18
Phacidium infestans, 18
Phoma blight, 11
Phoma spp., 11
 P. eupyrena, 11
Phomopsis blight, 9, 12
Phomopsis canker, 13
Phomopsis spp., 12, 13, 26
 P. juniperovora, 12
 P. lokoyae, 13
 P. macrospora, 26
 P. occulta, 13
Phyllactinia spp., 28
Phyllophaga spp., 52
Phyllosticta spp., 24
 P. minima, 24
Phytophthora root rot, 37
Phytophthora spp., 2, 33, 37, 38, 40
 P. cactorum, 37
 P. cinnamomi, 37
 P. citricola, 37
 P. cryptogea, 37
 P. dreschleri, 37
 P. gonapodyides, 37

P. megasperma, 37
P. pseudotsugae, 37
P. ramorum, 40
P. sansomeana, 37
Pikonema alaskensis, 22
Pine needle cast, 8
Pine tip moth, 21
Pitch canker, 14, 39
Pitch pine tip moth, 21
Plant bugs, 48
Plectrodera scalator, 29
Pleochaeta spp., 28
Ploioderma lethale, 8
Podosphaera spp., 28
Polyphylla spp., 52
Popillia japonica, 52
Poplar canker, 26
Poplar leaf rust, 27
Post-emergence damping-off, 14, 32, 33
Powdery mildew, 28
Pratylenchus spp., 36
Pre-emergence damping-off, 14, 32, 33
Pristiphora erichsonii, 22
Pseudocercospora spp., 9, 24
    P. fuliginosa, 24
    P. juniperi, 9
Pseudocneorrhinus bifasciatus, 51
Pythium root rot, 38
Pythium spp., 2, 33, 38
    P. aphanidermatum, 38
    P. irregulare, 38
    P. mamillatum, 38
    P. splendens, 38
    P. sylvaticum, 38
    P. ultimum, 38
Ramorum blight, 40
Rhizoctonia blight, 15
Rhizoctonia spp., 15, 33
Rhizopus spp., 39
Rhyacionia spp., 21
    R. bushnelli, 21
    R. frustrana, 21
    R. rigidana, 21
    R. subtropica, 21
Rhytisma spp., 24
    R. punctatum, 24
Root disease, 2, 31, 34, 36, 41, 54, 56
Root rot, 15, 31, 32, 36, 37, 38, 39
Root weevils, 51
Rust, 5, 6, 19, 27

Rust mites, 50
Salinity, 56
Salinity damage, 56
Salt damage, 56
Sarcotrochila spp., 18
Sawflies, 22
    European pine sawfly, 22
    hemlock sawfly, 22
    introduced pine sawfly, 22
    larch sawfly, 22
    lodgepole sawfly, 22
    redheaded pine sawfly, 22
    two-lined larch sawfly, 22
    white pine sawfly, 22
    yellowheaded spruce sawfly, 22
Scapteriscus spp., 47
    S. abbreviatus, 47
    S. borelli, 47
    S. vicinus, 47
Scarabaeidae, 52
Sciaridae, 45
Scirrhia acicola, 1
Scleroderris canker, 16
Sclerophoma pythiophyla, 12
Sclerotium bataticola, 31
Seed and cone insects, 49
Seed bugs, 49
Seed fungi, 39
Seed worms, 49
Septoria leaf spot, 24
Septoria spp., 24, 26
    S. alnifolia, 24
    S. musiva, 26
    S. populicola, 26
Serica spp., 52
Shieldbacked pine seed bug, 49
Shoot blight, 3, 10, 12, 13, 17, 23, 35, 40
Short-tailed crickets, 44
Shortwinged mole cricket, 47
Sirococcus shoot blight, 17
Sirococcus spp., 11, 17, 39
    S. conigenus, 17, 39
    S. piceae, 17
    S. strobilinus, 17
    S. tsugae, 17
Slash pine flower thrips, 49
Snow blight, 18
Snow mold, 18
Soluble salts, 41, 56
Southern mole cricket, 47

Sphaeropsis spp., 3, 11, 39
    S. sapinea, 3, 39
Sphaerotheca spp., 28
Spider mites, 41, 50
Spruce spider mites, 50
Streptopodium spp., 28
Sudden oak death, 40
Taphrina spp., 24
    T. populina, 24
Tawny mole cricket, 47
Taylorilygus apicalis, 48
Temperature extremes, 54
Tenthredinidae, 22
Tetranychidae, 50
Tetyra bipunctata, 49
Thanatephorus spp., 15
Tip blight, 10, 12, 15, 17, 40, 54
Tip moth, 21
Tortricidae, 21
Trichoderma spp., 35, 39
Trichodorus spp., 36
Trisetacus spp., 50
Tubakia dryina, 24
Tubakia leaf spot, 24
Tylenchorhynchus spp., 36
    T. claytoni, 36
    T. ewingi, 36
Uncinula spp., 28
Upper stem canker (see Fusarium), 34
Viruses, 41, 42
Weevils, 49, 51
    black vine weevil, 51
    Japanese weevil, 51
    obscure root weevil, 51
    pales weevil, 51
    pitch-eating weevil, 51
    root weevils, 51
    rough strawberry weevil, 51
    strawberry root weevil, 51
    woods weevil, 51
    yellow-poplar weevil, 51
Western conifer seed bug, 49
Western gall rust, 5
White grubs, 52
White pine blister rust, 19
White pine cone beetle, 49
Wind abrasion, 54
Winter burn, 54
Xiphinema spp., 36
    X. diversicaudatum, 36
Yellows, 41

Index of Host Plants

This index contains the common names and Latin names (in parentheses) of host trees, as listed in chapters. Chapter numbers follow the common and Latin names.

Alder (*Alnus spp.*) 30
    red (*A. rubra*) 24
Ash (*Fraxinus spp.*) 23
    Oregon (*F. latifolia*) 40
Bald cypress (*Taxodium distichum*) 9, 46
Beech (*Fagus spp.*) 40
Birch (*Betula spp.*) 23, 37
Black locust (*Robinia pseudoacacia*) 46
Black tupelo (*Nyssa sylvatica*) 46
Black walnut (*Juglans nigra*) 23, 28, 32, 37
Buckeye (*Aesculus spp.*) 28
California bay laurel (*Umbellularia californica*) 40
California nutmeg (*Torreya californica*) 40
Catalpa (*Catalpa spp.*) 23
Cedar, true (*Cedrus spp.*) 12, 17, 50
    eastern red (*Juniperus virginiana*) 9, 12, 44, 46, 47
    incense (*C. alocedrus decurreus*) 50
    northern white (*Thuja occidentalis*) 37
    western red (*Thuja plicata*) 13, 18, 49
Cherry (*Prunus spp.*) 24, 28
    chokecherry (*P. virginiana*) 24
Chestnut, American (*Castanea dentate*) 37
Common persimmon (*Diospyros virginiana*) 24
Cottonwood (*Populus spp.*) 28, 29, 30
    black (*P. trichocarpa*) 24
    eastern (*P. deltoids*) 29, 30
Cypress (*Cupressus spp.*) 9, 12, 50
    Arizona (*C. arizonica*) 46
Dogwood, flowering (*Cornus florida*) 28, 32, 39, 46
Douglas-fir (*Pseudotsuga menziesii*) 2, 11, 13, 18, 20, 34, 35, 37, 40, 48, 49, 50, 51
    coastal (*P. menziesii* var. *menziesii*) 13
    Rocky Mountain (var. *glauca*) 13
Elm (*Ulmus spp.*) 23, 28, 47
Eucalyptus (*Eucalyptus spp.*) 32
Fir, true (*Abies spp.*) 2, 3, 11, 18, 34, 37, 48, 49, 50, 51
    alpine (*A. lasiocarpa*) 18
    balsam (*A. balsamea*) 18
    California red (*A. magnifica*) 11, 18, 40
    European silver (*A. alba*) 18
    Fraser (*A. fraseri*) 37, 48, 54
    grand (*A. grandis*) 18, 40, 49
    noble (*A. procera*) 18, 20
    Pacific silver (*A. amabilis*) 18, 22
    red (*A. magnifica*) 11, 18, 40
    subalpine (*A. lasiocarpa*) 18
    white (*A. concolor*) 11, 18, 40

Index of Host Trees

Hemlock (*Tsuga* spp.) 17, 22, 35, 37, 50, 51
    eastern (*T. canadensis*) 18
    mountain (*T. mertensiana*) 17, 18
    western (*T. heterophylla*) 13, 18, 40, 49
Hickory (*Carya* spp.) 23, 28, 49
Horse chestnut (*Aesculus hippocastamon*) 40
Juniper (*Juniperus* spp.) 9, 12, 18, 50
    Alligator (*J. deppeana*) 18
    Creeping (*J. horizontalis*) 18
    Rocky Mountain (*J. scopulorum*) 9, 18
Larch (*Larix* spp.) 3, 7, 20, 27, 34, 35, 37, 40, 48, 50
    European (*L. decidua*) 7
    hybrid (*L. eurolepis*) 7
    Japanese (*L. kaempferi*) 7, 40
    Siberian (*L. sibirica*) 7
    tamarack (*L. laricina*) 22
    western (*L. occidentalis*) 2, 7, 13, 22, 37, 49
Linden (*Tilia* spp.) 23
Maple (*Acer* spp.) 23, 24, 28, 40, 41, 47
    bigleaf (*A. macrophyllum*) 24, 40
    Norway (*A. platanoides*) 28
    red (*A. rubrum*) 24, 41
Oak (*Quercus* spp.) 5, 6, 23, 24, 28, 36, 37, 40, 49, 50
    black (*Q. velutina*) 23
    bur (*Q. macrocarpa*) 5, 23
    California black (*Q. kelloggii*) 40
    canyon live (*Q. chrysolepis*) 40
    cherrybark (*Q. pagoda*) 32
    coast live (*Q. agrifolia*) 40
    red, northern (*Q. rubra*) 23, 32
    Shreve's (*Q. parvula* var. *shrevei*) 40
    Southern red (*Q. falcate*) 6
    water (*Q. nigra*) 6
    white (*Q. alba*) 23, 54
Pacific madrone (*Arbutus menziesii*) 40
Pecan (*Carya illinoinensis*) 23, 53
Pine (*Pinus* spp.) 1–6, 8, 10, 14–19, 21, 31, 34, 36, 37, 39, 41, 44, 48–50, 52, 54, 55
    Aleppo, (*P. halepensis*) 5, 8
    Austrian (*P. nigra*) 3, 4, 8, 16
    bishop (*P. muricata*) 5, 14
    bristlecone (*P. aristata*) 18, 19
    Canary Island (*P. canariensis*) 5
    Coulter (*P. coulteri*) 5
    Cuban (*P. cubensis*) 8
    eastern white (*P. strobes*) 1, 8, 10, 16, 18, 19, 21, 22, 32, 49, 54
    foxtail (*P. balfouriana*) 18
    gray (*P. sbiniana*) 5, 8
    jack (*P. banksiana*) 3, 5, 16, 17, 22
    Japanese black (*P. thunbergii*) 8
    Jeffrey (*P. jeffreyi*) 5, 8, 18
    knobcone (*P. attenuata*) 5, 14

Index of Host Trees

limber (*P. flexilis*) 8, 18, 19
loblolly (*P. taeda*) 1, 6, 8, 14, 15, 21, 22, 36, 46, 48, 49, 54, 55
lodgepole (*P. contorta*) 4, 5, 8, 11, 16, 17, 18, 22, 49
longleaf (*P. palustris*) 1, 6, 14, 15, 21, 22, 32, 39, 49
Monterey (*P. radiata*) 3, 4, 5, 8, 14, 21, 39
mugo (*P. mugo*) 3, 5, 8, 11, 18
pitch (*P. rigida*) 1, 6, 8, 21
pond (*P. serotina*) 1, 6, 8
ponderosa (*P. ponderosa*) 3, 4, 5, 8, 11, 18, 21, 49, 50
red (*P. resinosa*) 3, 8, 16, 17, 21, 22, 32, 37
rough-barked Mexican (*P. montezumae*) 8
sand (*P. clausa*) 8, 14, 46, 48
Scots (*P. sylvestris*) 1, 3, 5, 8, 16, 21, 22, 35
shortleaf (*P. echinata*) 1, 5, 6, 8, 14, 21, 22, 49
slash (*P. elliottii*) 1, 6, 8, 14, 15, 21, 22, 31, 36, 39, 44, 46, 49
Sonderegger (*P. palustris* x *P. taeda*) 1
Southern yellow (*P. echinata*, *P. elliottii*,
    P. palustris, and *P. taeda*)............................47
southwestern white (*P. strobiformis*) 19
spruce (*P. glabra*) 1, 8, 14
sugar (*P. lambertiana*) 18, 19, 49
Table Mountain (*P. pungens*) 8
Virginia (*P. virginiana*) 1, 5, 8, 14, 21
western white (*P. monticola*) 2, 4, 18, 19, 49
whitebark (*P. albicaulis*) 2, 18, 19
Poplar (*Populus* spp.)..25, 26, 27, 29, 30, 48
    balsam (*P. balsamifera*).........................25
    white (*P. alba*) 25
Quaking aspen (*Populus tremuloides*)...........................25
Redbud (*Cercis Canadensis*) 32
Redwood (*Sequoia* spp. and *Sequoiadendron* spp.)........35, 40, 50
    coast (*Sequoia sempervirens*)................................40
    giant sequoia (*Sequoiadendron giganteum*) 9, 35
Spruce (*Picea* spp.)...3, 17, 18, 20, 22, 34, 35, 37, 48, 49, 50, 51, 56
    black (*P. mariana*)..................................18, 32
    Colorado blue (*P. pungens*)...................17, 18
    Engelmann (*P. engelmannii*).................11, 18, 35
    Norway (*P. abies*)...................................18
    red (*P. rubens*) 18, 32
    Sitka (*P. sitchensis*) 13, 17, 18
    white (*P. glauca*) 18, 32
Sweetgum (*Liquidambar styraciflua*) 32, 47
Sycamore (*Platanus occidentalis*) 23, 28, 46, 50
Tanoak (*Notholithocarpus densiflorus*)..........................40
Willow (*Salix* spp.) 24, 29, 30
Yellow-poplar (*Liriodendron tulipifera*) 23, 28, 32, 47
Yew (*Taxas* spp.) 18, 37, 40
    European (*T. baccata*)............................40
    Pacific (*T. brevifolia*) 18, 40

Glossary

abiotic. Of or pertaining to the nonliving; inanimate.

abiotic disease. Disease resulting from nonliving agents.

abscise. Leaves or buds are shed following the formation of a separation layer of cells (abscission layer) in response to injury, disease or senescence.

acervulus (pl., acervuli). A small cushionlike asexual fruiting body, without a covering of fungus tissue, which produces conidia spores in a moist mass on the host tissue.

adulticide. A chemical that kills only adult insects.

aeciospore. A nonrepeating, asexual spore, usually orange or yellow, produced by some rust fungi. Aeciospores are incapable of infecting the host on which it is produced.

anaerobic. Process or microorganism that occurs in the absence of oxygen.

alpha-spore. A fertile asexual spore (conidia) of fungi in the Diaporthaceae family. Spores are fusoid to oblong and can be produced with beta-spores.

anamorph. A mitotically reproducing form of a fungus, usually an asexual state of an ascomycete or basidiomycete.

anthracnose. A leaf, twig or fruit disease characterized by necrotic spots or lesions and generally caused by fungi that produce spores in an acervulus.

apothecium (pl., apothecia). A sexual fruiting body that are shaped like a cup, saucer, or wineglass that produces ascospores.

appressed. Flattened or pressed closely against a surface.

arbuscular mycorrhiza. An association between a plant root and a mycorrhizal fungus; the fungus invades the cortical cells of the root and increases the plants nutrient and water uptake in exchange for carbon.

ascomycete. A large group of fungi characterized by the formation of spores, usually eight in number, in a saclike structure called an ascus.

ascus (pl., asci). A saclike cell of an ascomycete fungus, in which ascospores are produced.

asymptomatic. Without symptoms.

ascospore. A spore produced in the sexual fruiting body of an ascomycete.

asexual state. A vegetative state or a reproductive state in the life cycle of a fungus in which nuclear fusion is absent and in which reproductive spores are produced by mitosis or simple nuclear division.

Baermann funnel. A funnel with rubber tubing; used for isolating nematodes from the soil.

basidiomycete. Member of a large group of fungi characterized by the production of external spores, usually four, on a basidium.

basidiospore. A spore produced by the sexual state of the basidiomycetes.

basidium (pl. basidia). A cell, usually terminal, in which nuclear fusion and meiosis occur and on which haploid spores (usually four) are produced.

beta-spore. An infertile spore produced together with alpha-spores by certain fungi in diaporthaceae. Beta-spores are long, slender and usually curved or bent.

binucleate. A cell with two nuclei.

biological control. The use of natural enemies to reduce or mitigate pests and their damage.

biotic. Of or pertaining to a living organism.

blastospore. A fungal spore that is produced by budding.

blight. A plant disease that causes rapid death or dieback of a plant or part of a plant.

blotch. A large, irregular necrotic area on a leaf or fruit.

borer. Insect or insect larva that tunnels within the wood of trees.

brood. All individuals that hatch at approximately the same time from eggs laid by one set of parents.

broom. An abnormally dense mass of host branches and foliage, in which the typical growth pattern is replaced by a disordered cluster of foliage at the branch tips.

bug. Any insect of the order Hemiptera characterized by sucking mouthparts and two pairs of wings.

callus. A protective tissue of thin walled cells developed around the edges of wounds or necrotic lesions.

cambium. The layer of cells that lies between and gives rise by cell division to the secondary xylem(wood) and the secondary phloem (inner bark).

canker. A well defined, relatively localized necrotic lesion primarily of the bark and cambium.

casting. Premature loss of leaves or needles.

caterpillar. The elongated wormlike larva of a butterfly or moth.

causal agent. An organism, such as a fungus, bacterium, or virus, which produces a disease.

chasmothecia. A spherical ascocarp that produces ascospores.

chlamydospore. A thick-walled asexual resting spore formed directly from hyphal cells; typically formed by many soilborne fungi.

chlorosis. An abnormal yellowing of the foliage.

chlorotic. Abnormal yellowish.

clone. All descendents derived from a single individual by asexual reproduction, or parthenogenesis.

cocoon. An envelope, often largely of silk, which an insect larva forms around itself as protection for the pupal stage.

coelomycetes. A fungus that produces condia within a fruiting structure. Are anamorphs of ascomycetes, or have no known sexual state.

coenocytic. A mycelium where the hyphae lack septa.

Glossary

collar rot. Rotting of the stem at or near the ground or soil line.

colonize. To establish an infection within a host or part of a host.

complex diseases. A disease caused by the interaction of two or more pathogens.

conidiophore. A specialized hypha that produces asexual spores called conidia.

conidium (pl., conidia). An asexual spore of a fungus typically produced at the end of a specialized hypha called a conidiophore.

cortex. The primary tissue found between the epidermis and the stele of a stem or root.

cotyledon. The food-storing protion of an embryo; also known as the seed leaf.

cover crop. A crop, natural or introduced, that is grown alternately with the main crop. Used to prevent erosion and to improve soil characteristics.

cull. A seedling that is rejected because it does not meet certain specifications.

cultural practices. A general term for those routine nursery operations required to help seedling growth, such as irrigation, weeding, plowing, etc.

cuneate. Wedge-shaped and more narrow at one end.

cuticle. A thin, waxy layer on the outer wall of epidermal cells.

damping-off. A disease of germinating seed and seedlings characterized by mortality prior to emergence (preemergence) or the collapse of the seedling stem at ground level immediately after germination (postemergence).

decay. The decomposition of plant tissue by fungi and other microorganisms.

decline. The gradual reduction in health and vigor of a plant.

desiccation. An excessive loss of moisture; drying out.

dieback. The progressive dying of the stem and branches from the tip downward.

disease. An unfavorable change in the function or form of a plant, caused by pathogenic agents, environmental factors, or a complex of factors.

disease cycle. The chain of events involved in disease development, including the stage of development of the pathogen and effect of the diseases on the host.

distal. Near or toward the free end of any appendage; that part of a segment farthest from the body.

DNA. Deoxyribonucleic acid: a nucleic acid that makes up chromosomes; contains genetic information.

echinulate. Having many small spines.

ectomycorrhizae. A type of mycorrhizal association in which the fungal component grows between and/or external to the cortical cells of the plant root.

ectoparasite. A parasite that occurs and feeds outside its host.

ELISA. Enzyme-linked immunosorbent assay used to identify a specific protein, especially an antigen or antibody; used in diagnostic tests.

emergence. The escape of a winged adult from its cocoon, pupal case or nymph.

endemic. Native to the country or region; existing at low stable population levels.

endomycorrhizae. A type of mycorrhizal association in which the fungal component invades the cortical cells of the root (syn. arbuscular mycorrhizae).

endoparasite. A parasite that lives within its host.

entomophagous. Feeding on insects.

epidemic. Pertaining to a disease that has built up rapidly and reached injurious levels.

epidemiology. Factors influencing initiation, development and spread of disease.

epidermis. The outermost layer of cells on the primary plant body.

ellipsoid. Shape of an object with cross-sections that are either oval or circular.

exotic. Introduced from another country or area.

exudate. Matter that oozes out or is secreted.

facultative parasite. An organism that is normally saprophytic but that is capable of living as a parasite.

fallow. Cultivated land allowed to lie idle or unplanted during the growing season.

field capacity. The maximum amount of water a soil can hold against the force of gravity.

filiform. Long and slender; threadlike.

flaccid. Limp or nonturgid.

flag. On a living plant, a conspicuous dead branch with the foliage attached.

frass. Insect excrement and refuse.

fruiting body. Fungal reproductive structures that produce spores.

foci. Central points from which a disease develops or in which it localizes.

forma specialis (pl., formae speciales). An intraspecific population of plant pathogenic species distinguished by host preference, but not morphologically distinguishable from other members of the species.

fumigation. Application of vapor or gas, especially for the purpose of disinfecting or destroying pests.

fungus (pl., Fungi). An undifferentiated plant that lacks chlorophyll and conducive tissues.

Fungi Imperfecti. A group of miscellaneous fungi that lack a known sexual state.

fusiform. Spindle-shaped; tapering toward each end.

gall. An abnormal swelling on a plant caused by certain fungi, bacteria, insects, or nematodes.

gallery. Passage made in wood by an insect.

gametophyte. Stage in the life cycle of an organism bearing or producing the sexually reproductive structures.

geniculate. Bent abruptly at an angle, like a knee.

genotype. The genetic composition of an organism.

germ tube. The hypha produced by a germinated fungus spore.

girdle. To destroy or remove the tissue in a ring around a stem, branch, or root, casing a disruption of the xylem and the phloem.

globose. In the shape of a globe or ball.

glomalin. A sticky glycoprotein produced by arbuscular mycorrhizae.

host. The plant or animal that affords nourishment to a parasite.

host range. All hosts that a particular pathogen or insect attacks.

host-specific. A term used to describe pathogens or insects that attack only certain species of hosts.

hyaline. Transparent; having no color.

hypha (pl., hyphae). One of the filamentous threads that make up the fungus body.

hypocotyl. That part of the axis of a developing embryo just below the attachment of the cotyledons.

imperfect state. The anamorph or asexual stage in the life cycle of a fungus.

incite. To cause a disease.

infect. To invade and cause a disease.

infest. To attack, inhabit, or populate a thing or place.

inoculate. To place a pathogen on or in a host.

inoculum. The spores, mycelium, sclerotia, or other propagules of a pathogen that initially infects a host.

instar. The period or stage between molts in larvae, numbered to designate the various periods. The first instar, for example, is the period between the egg and the first molt.

intercellular. Occurs or grows between the cells.

intracellular. Occurs or grows within the cells.

lancet. Any piercing mouth structure.

larva (pl., larvae). The immature stage between the egg and pupa of an insect, which undergoes complete metamorphosis (egg, larva, pupa, and adult).

latent infection. An established infection without visual symptoms.

leaf spot. A leaf disease characterized by distinct lesions.

lesion. A well-defined, localized area of diseases tissue.

longicorn. An insect having antennae as or longer than the body; specifically belonging to the family Cerambycidae.

macroconidia. The larger of two types of conidia produced by certain fungi, such as *Fusarium* species.

maggot. A white to creamy larvae without legs or a well developed head capsule; in the order Diptera (flies).

mandible. Anterior pair of jaws on an insect.

microconidia. The smaller of the two types of conidia produced by certain fungi.

micron. A unit of measurement; 1/1,000 millimeter, 1/25,400 inch.

microsclerotium. Small, dense aggregate of darkly pigmented, thick-walled hyphal cells, which serve as resting structures.

mildew. A plant disease characterized by a coating of mycelium or spores or both on the surface of the affected parts.

moribund. At the point of mortality.

morphology. The external form and structure of organisms.

motile. Having the power of motion.

mycelium (pl., mycelia). A mass of hyphae that forms the vegetative, filamentous body of a fungus.

mycoflora. The fungi characteristic of a region or environment.

mycoplasma. A wall-less prokaryotic microorganism.

mycorrhiza (pl., mycorrhizae). A symbiotic association between a fungus and the roots of higher plants that aids in the uptake of nutrients by the plant.

necrosis. Death of plant cells, usually resulting in darkening of the tissue.

needle cast. A disease of conifer needles that usually results in premature needle drop.

nymph. Immature stage of certain insects having incomplete metamorphosis, (egg, nymph, and adult).

obligate parasite. An organism that can survive only in living tissue.

oospore. The sexual resting spore produced by certain fungi in the class oomycetes.

oviposit. The act of depositing eggs.

ovipositor. An external egg-laying apparatus of female insects.

parthenogenesis. Reproduction by growth of egg cells without male fertilization.

pathogen. An organism that causes a disease.

pathogenic. Capable of causing a disease.

parasite. An organism living on or nourished by another living organism.

perfect state. The teleomorph or sexual stage in the life cycle of a fungus.

periderm. The outer, protective layer of stems, consisting of the phellogen, phellum, and phelloderm.

Glossary

perithecium (pl., perithecia). A closed, flask-like sexual fruiting body in which ascospores are produced. Formed by certain ascomycetes.

phellum. The suberized tissue produced by the cork cambium in the bark.

phelloderm. Secondary tissue produced by and to the inside of the cork cambium.

phialide. A cell that develops one or more open ends from which a succession of conidia emerge without increasing the length of the cell itself.

phloem. The tissues of the inner bark responsible for the transport of photosynthates.

phytotoxic. A chemical that is toxic to plants.

pionnotal. A degenerate cultural variant; a flat slimy culture with abundant condidia.

polyphialide. A phialide with more than one open end.

proleg. A fleshy abdominal leg of certain insect larvae; prolegs occur in pairs.

pseudothecium (pl., pseudothecia). Asci produced in cavities (locules) located within a stroma (usually black); resembles a perithecium.

pupa (pl., pupae). The resting, inactive stage of an insect between larva and adult.

pustule. Small blisters created by a fungus that mature into fruiting structures.

pycnidium (pl., pycnidia). A fungal fruiting body, typically flask shaped, in which asexual spores are produced.

pycnidiospore. An asexual spore produced in a pycnidium.

pycniospore. A specialized spore, produced in a pycnium by rust fungi, that functions as male gamete. Synonym: spermatium.

pycnium (pl., pycnia). A structure developed by rust fungi that produces tiny, one-celled spores which function as male gametes.

resistant. Able to withstand, without serious injury, attack by an organism or damage by a nonliving agency, but not immune from such attack.

root collar. The area where the major roots join the together with the base of the stem; root crown.

rot. See Decay.

rust. A disease caused by certain fungi in the Basidiomycetes and usually characterized by the production of large numbers of reddish (rusty) spores on foliage, branches, or stems.

saprophyte. An organism that uses dead organic material as food.

sclerotium (pl., sclerotia). A firm, frequently rounded, multicellular resting structure produced by fungi.

scorch. The sudden browning of large, indefinite areas on a leaf; caused by infection, chemical injury, or unfavorable weather conditions.

senescent. Late stage of a plants lifecycle, eventually leading to mortality.

septate. Having cross walls, or septa, that divide hyphae or spores into a number of separate cells.

septum (pl., septa). The cross walls that divide a hypha or spore into two or more distinct cells.

sexual state. The state in the life cycle of a fungus in which spores are produced after sexual fusion. Synonym: perfect state.

sign. Vegetative or fruiting structures of the casual organism on a diseased plant. Along with symptoms, signs are used to diagnose cause(s) of disease.

sporangium (pl., sporangia). A cell that contains one or more asexual spores.

spore. The reproductive structure of the fungi and other lower plants.

sporodochium (pl., sporodochia). A conidial fruiting body in which the spore mass is supported by a cushionlike mass of short conidiophores.

sporulate. To produce spores.

stage. Any definite period in the development of an insect, for example, egg and larva.

state. One spore type produced by a fungus, which produces two or more spore types during its life cycle. Sometimes referred to as stage.

stoma (pl., stomata). A pore in the leaf epidermis, surrounded by two guard cells, leading into an intercellular space within the plant.

stool. A plant from which offsets may be taken or with several stems arising together; a clump of roots or root stocks that may be used in propagation.

stroma (pl., stromata). A cushionlike body on or in which fungus fruiting bodies are formed.

stylet. A hollow protractible spear used to puncture plant or animal prey.

sublethal infection. An infection that does not result in death of the host.

susceptible. Unable to withstand, without serious injury, attack by an organism or damage by a nonliving agency.

symbiosis. A mutually beneficial association of two or more different types of organisms.

symptom. The visual evidence of disturbance in the normal development and function of a host plant, for example, chlorosis, necrosis, galls, and stunting.

systemic. Affecting or distributed throughout the whole plant.

taproot. The primary descending root of a plant from which the secondary, or lateral, roots branch.

Glossary

teliospore. The sexual spore state of a rust fungus from which the basidium and basidiospores arise upon germination.

telium (pl., telia). The structure in rust fungi that gives rise to teliospores.

tendrils. Mass of spores in a gelatinous matrix, which oozes from a fruiting body in long curling strings or hornlike projections.

thorax. The second or intermediate region of the insect body bearing the true legs and wings.

translocation. The transfer of food materials or metabolites within a plant.

truncate. Cut off, a straight edged base.

urediospore. A binucleate, asexual spore produced by some rust fungi, which is capable of reinfecting the host on which it is formed. Sometimes called urediospore.

uredinium (pl., uredinia). The structure in rust fungi that gives rise to urediniospores. Sometimes called uredium.

vascular. Plant tissue that conducts water, food, and hormones within the plant.

vegetative. Concerned with growth and development, as distinguished from reproductive functions.

vesicle. A bladderlike sac, the swollen apex of a conidiophore or hypha.

viable. Capable of becoming normally active.

wilt. A type of plant disease characterized by the sudden wilting and collapse of the succulent parts of affected plants.

virulent. Capable of causing severe disease; strongly pathogenic.

sylem. The woody water-conducting tissues of stems and roots.

zoospore. A motile free-swimming spore produced by the water molds.

This publication reports research with pesticides. It does not contain recommendations for their use, nor does it imply that the uses discussed here have been registered. All uses of pesticides must be registered by appropriate State and/or Federal agencies before they can be recommended.

CAUTION: Pesticides can be injurious to humans, domestic animals, desirable plants, and fish or other wildlife—if they are not handled or applied properly. Use all pesticides selectively and carefully. Follow recommended practices for the disposal of surplus pesticides and pesticide coontainers.

The mention of products and companies by name does not constitute endorsement by the U.S. Department of Agriculture, nor does it imply approval of a product to the exclusion of others that may also be suitable.

This publication reports research with pesticides. It does not contain recommendations for their use, nor does it imply that the uses discussed here have been registered. All uses of pesticides must be registered by appropriate State and/or Federal agencies before they can be recommended.

CAUTION: Pesticides can be injurious to humans, domestic animals, desirable plants, and fish or other wildlife—if they are not handled or applied properly. Use all pesticides selectively and carefully. Follow recommended practices for the disposal of surplus pesticides and pesticide coontainers.

The mention of products and companies by name does not constitute endorsement by the U.S. Department of Agriculture, nor does it imply approval of a product to the exclusion of others that may also be suitable.

www.ingramcontent.com/pod-product-compliance
Ingram Content Group UK Ltd.
Pitfield, Milton Keynes, MK11 3LW, UK
UKHW051652180426
11946UKWH00005B/124